The New
Cosmic Onion

Quarks and
the Nature of the Universe

Frank Close
University of Oxford,
England

Taylor & Francis
Taylor & Francis Group
New York   London

Taylor & Francis is an imprint of the
Taylor & Francis Group, an informa business

CRC Press
Taylor & Francis Group
6000 Broken Sound Parkway NW, Suite 300
Boca Raton, FL 33487-2742

© 2007 by Taylor & Francis Group, LLC
CRC Press is an imprint of Taylor & Francis Group, an Informa business

No claim to original U.S. Government works
Printed in the United States of America on acid-free paper
10 9 8 7 6 5 4 3 2 1

International Standard Book Number-10: 1-58488-798-2 (Softcover)
International Standard Book Number-13: 978-1-58488-798-0 (Softcover)

This book contains information obtained from authentic and highly regarded sources. Reprinted material is quoted with permission, and sources are indicated. A wide variety of references are listed. Reasonable efforts have been made to publish reliable data and information, but the author and the publisher cannot assume responsibility for the validity of all materials or for the consequences of their use.

Library of Congress Cataloging-in-Publication Data

Close, Frank.
    The new cosmic onion : quarks and the nature of the universe / Frank Close.
        p. cm.
    Rev. ed. of: The cosmic onion, 1983.
    Includes index.
    ISBN 1-58488-798-2 (alk. paper)
    1. Quarks. 2. Cosmology. I. Title.

QC793.5.Q252C57 2007
539.7'2167--dc22   2006050482

Visit the Taylor & Francis Web site at
http://www.taylorandfrancis.com

and the CRC Press Web site at
http://www.crcpress.com

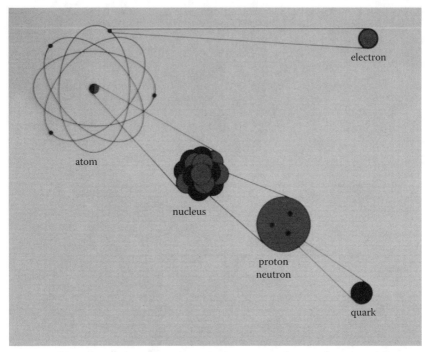

Typical sizes: 1/100,000,000 cm   1/1,000,000,000,000 cm   1/10,000,000,000,000 cm
1/100,000,000,000,000 cm (or less)

These minute distances can be more easily written as 10^{-8} cm for atoms, 10^{-12} cm for nucleus, and 10^{-13} cm for a nuclear particle. Electrons and quarks are the varieties of matter that exist as distances of less than 10^{-14} cm. These are the shortest distances that present technology can probe. (Source: CERN.)

Foreword

Why are scientists spending $10 billion to build the Large Hadron Collider (LHC)? This question is being increasingly asked as this most ambitious project in particle physics prepares for its grand opening in 2007. In an attempt to provide some of the answers for a public who know of science but who are not necessarily professional scientists, I decided to produce this new version of *The Cosmic Onion*.

The original version first appeared in 1983: a time when popular accounts of particle physics were almost non-existent. Its title became a brand in its own right, being widely adopted as a metaphor for the structure of matter consisting of ever deeper layers: from galaxies of stars, to atoms, and the basic seeds — quarks. Today, by contrast, there is a vast literature on popular physics, increasingly focussed on exciting but highly speculative ideas such as superstring theory, higher dimensions, and parallel universes. Sometimes it is difficult for the public to distinguish between what is science fact and science fiction or, in the words of Bill Bryson,[1] "legitimately weird or outright crackpot; twaddle, a work of genius or a hoax." I restricted the original *Cosmic Onion* to established concepts that, while they once might have been classified as weird genius, had matured to established conservatism that I felt would last: a quarter of a century later, I am pleased to say that this has turned out to be the case. But in that period a number of things have happened that have alerted me to the passage of time and the need for *The New Cosmic Onion*.

First, there was the personal shock when, about 10 years ago, undergraduates started telling me that *The Cosmic Onion* had been their first childhood introduction to the fascinating subject of particle physics. In the past 5 years, these childhood experiences were being recalled by post-doctoral research fellows; and when a tenured physicist told me as much, I felt it was time to start writing before it was too late. On the scientific front, new discoveries have radically changed our perception of matter and the universe, many of which were undreamed of in 1983. These all highlighted how many years had passed since I wrote the original: as the world of particle physics prepared for the LHC, I decided it was high time to bring *The Cosmic Onion* up to date.

Originally written as a popular description of the nature of matter and the forces that control the universe, it gained a much wider following. It was used as a recommended background book in universities and schools; Sir John Kendrew's national committee reviewing Britain's role in world high-energy physics and at least one U.K. government minister of science used it as a brief; it led to the author's televised Royal Institution Christmas Lectures "The Cosmic Onion" in 1993, as well as inspiring a generation of students to take up science.

By focusing on concepts that were in my opinion established, I am pleased that almost nothing of that original *Cosmic Onion* has proved to be so highly conjectural that it has been overthrown. Many new facts have come to be known, including several

---

[1] Bryon, B., *A Short History of Nearly Everything*, Broadway Books, 2003.

that were not anticipated, and much of the material in the latter part of this new version bears testimony to that.

Discoveries and precision measurements during the past 10 years have established Electroweak Theory as a law of nature. The discovery of the top quark had been expected, but it was a surprise that it turned out to be so massive, some thirty times greater than its sibling, the bottom quark. A question for the future is whether this makes the top quark the odd one out or whether it is the only "normal" one, existing on the scale of electroweak symmetry breaking, while all the other fundamental particles are anomalously light. In any event, the top quark is utterly different and when the LHC enables it to be studied in detail for the first time, further surprises could ensue. The bottom quark turned out to be surprisingly stable, which has opened a new window into the mystery of the asymmetry between matter and antimatter. We have looked into the heart of the Sun and even a supernova by means of neutrinos; these have shown that neutrinos are not simply massless particles that travel at light speed, and in turn this is opening up new research lines that were not on the theorists' agenda in 1983. Ten years of precision data from LEP at CERN in Geneva have revealed the subtle quantum mechanical influence that the Higgs field has on the vacuum and established the criteria by which the Higgs Boson may be produced. In the cosmological arena we are now coming to the opinion that the universe is "flat," with appreciable dark matter (about which some were beginning to speculate 20 years ago) and also dark energy (which was utterly unexpected). Quarks and leptons have been shown to be structureless particles to distances as small as 10^{-19} m, which is as small relative to a proton as the proton is to the dimensions of a hydrogen atom.

While these advances have been delivered by experiment, there were also theoretical concepts, such as superstrings, that in 1983 were only just about to emerge, and which are now a major mathematical research area. It is possible that this will eventually prove to be the long-sought Theory of Everything; and to read some popular science, or watch it on television, you might have the impression that it is established lore. However, there is at present no clear evidence to show that it has anything to do with the physics that has been revealed by experiment. Time and experiment will tell, and that is for the future. For the present, that is why *The New Cosmic Onion* comes with no strings attached.

The New Cosmic Onion contains the best from the original, thoroughly revised and updated, plus extensive new material that explains the scientific challenges at the start of the 21st century. New accelerators are being built that will show how the seeds of matter were created when our universe was less than a billionth of a second old. The discoveries in this century promise to be no less revolutionary than in the last. The hope is that *The New Cosmic Onion* will provide the explanations that students, opinion formers, and intelligent citizens need if they are to understand how science has come to this frontier, and where we think it is headed in the immediate future. As in the original, I have selected the content that I believe will last, and have avoided flights of fancy that might not survive the test of time.

I am grateful to the many people who, having read the original, wrote to me with corrections, questions, and suggestions. In particular, I am grateful to my physics students at Exeter College, Oxford, who have read the original as part of their summer studies and identified points that needed better explanations. Some readers' suggestions are mutually exclusive, so the new version will not be to everyone's taste. Nonetheless, I anticipate and welcome comments, in the hope that once the discoveries start to flow from the LHC, revisions will be able to take such comments into account. However, one result of writing such books is that the author receives countless letters with news of the writer's proof that Einstein was wrong, or of them having stumbled on some code of numbers that explains the masses of the particles in terms of π or e. Science advances by novel ideas, certainly, but it is essential that they not only fit known facts, but also make some testable prediction. It is experiment that decides what is true or false; that is what I have used as a paradigm in deciding what to include in and what to omit from this book.

I am grateful to my students for their penetrating questions, and to my long-suffering family and friends who have been used as sounding boards for potential answers. I am grateful to Anne and Stuart Taylor for the elegant writing desk at their Norfolk cottage, and to the Norfolk Wildlife Trust for providing such well-situated bird hides that enabled my wife to spend hours watching nature through her telescopes, while I passed the time waiting for the Bittern (that never showed itself) by writing about Nature as revealed by the ultimate microscopes.

Frank Close
Oxford, 2006

Suggestions for Further Reading

This is not intended as a comprehensive guide, but rather as a means of extending or deepening your reading on particle physics.

For the experimental aspects of particle physics, together with many images of facilities, detectors, and particle trails, see *The Particle Odyssey*, Frank Close, Michael Marten, and Christine Sutton (Oxford University Press, 2003) or its predecessor, *The Particle Explosion*. A brief summary suitable for undergraduate background reading is *Particle Physics — A Very Short Introduction*, Frank Close (Oxford University Press, 2004).

A detailed technical introduction, which is excellent for serious students, is *Nuclear and Particle Physics*, W.S.C. Williams (Oxford University Press, 1994).

A classic review of modern ideas in theoretical particle physics is *Dreams of a Final Theory* by Steven Weinberg (Pantheon, 1992; Vintage, 1993). Two Nobel Laureates who shared in the creation of the birth of electroweak theory have written semi-popular books on the subject: *In Search of the Ultimate Building Blocks* by G. 't Hooft (Cambridge University Press, 1996) and *Facts and Mysteries in Elementary Particle Physics* by M. Veltman (World Scientific, 2003).

Asymmetry and the ideas of spontaneous symmetry breaking that underpin the Higgs mechanism are described in *Lucifer's Legacy* by Frank Close (Oxford University Press, 2000), and by Henning Genz in *Nothingness* (Perseus Books, 1999).

Brian Greene's *The Elegant Universe, Superstrings, Hidden Dimensions* and the *Quest for Ultimate Reality* (Jonathan Cape, 1999) and Gordon Kane's *The Particle Garden: Our Universe as Understood by Particle Physicists* will take you to the ideas on superstrings and other more speculative areas that are not covered in *The New Cosmic Onion*. For a contrary viewpoint about string theory, which is perhaps nearer to the thesis of the present book, try *Not Even Wrong* by P. Woit (Jonathan Cape, 2006).

I have, as far as possible, tried to avoid the mysteries of quantum mechanics. For those who want a popular and informed introduction, read *The New Quantum Universe* by Tony Hey and Patrick Walters (Cambridge University Press, 2003). A more formal but accessible introduction to the basic concepts is *Quantum Mechanics — A Very Short Introduction*, John Polkinghorne (Oxford University Press). A history of the early days of quantum mechanics, and the birth of particle physics is *Inward Bound* by Abraham Pais (Oxford University Press, 1986).

A classic nontechnical description of the aftermath of the Big Bang is *The First Three Minutes* by Steven Weinberg (Andre Deutsch, 1977; Basic Books, 1993).

For more on the history of quarks, read *The Hunting of the Quark* by Michael Riordan (Simon and Schuster, 1987) and for the quark-related work of Murray Gell Mann, *Strange Beauty* by G. Johnson (Vintage, 2000). For more general history of 20th century particle physics, *The Particle Century* is a collection of articles edited by

Gordon Fraser (Institute of Physics, 1998), and the glorious days of the birth of nuclear physics are described in *The Fly in the Cathedral* by Brian Cathcart (Penguin-Viking, 2004). The most complete and a highly readable history of 20th century nuclear physics is the Pulitzer Prize winning *The Making of the Atomic Bomb* by Richard Rhodes (Simon and Schuster, 1986; Penguin, 1988).

Table of Contents

1 The Nature of the Universe

Nearly half a century has passed since the day in 1965 when Arnold Penzias and Robert Wilson, two American astronomers, unwittingly discovered the fading whispers of the Big Bang. For months they had been using a radio antenna in Holmdel, New Jersey, to detect signals coming from outer space but were frustrated by a background noise similar to the "static" that can interfere with reception of a concert broadcast on a traditional (analogue) radio. The hiss was constant, unvarying, and seemed to come from all directions. At first they thought that pigeon droppings were causing it, but after cleaning the antenna, the noise continued.

They had no idea what was going on. Some 30 years earlier, another American astronomer, Edwin Hubble, had discovered that the galaxies of stars are rushing away from one another: the universe is expanding. Put Hubble's expanding universe with Penzias and Wilson's radio noise and you have the seeds of our modern understanding of creation: the Big Bang theory of genesis.

As a result of many experiments and more refined measurements than were possible in Penzias and Wilson's day, we now have a confident story of how our material universe originated and developed into its present form. We now view the universe as having three types of ingredients: (1) matter, such as us, the Earth, Sun and stars; (2) antimatter, which is a favourite of science fiction but, as we shall see, is a real "mirror image" of matter; and (3) the cold "background radiation." This all fits with the idea that the universe erupted in a hot Big Bang, spewing matter and radiation outwards from a hot fireball. This is the source of the expanding universe that Hubble first observed. As the universe expanded, the radiation cooled. Hot bodies glow white, less hot ones yellow or red, and at room temperature they emit infra-red radiation: you can feel the warmth of your own body even though you are not shining, but an infra-red camera can take your picture. Cooler bodies emit radio waves, and from the wavelength you can determine the temperature. The background noise that Penzias and Wilson had chanced upon was a radio signal that was strongest around a wavelength of 1 mm, in the microwave part of the electromagnetic spectrum, implying that its source is very cold: $-270°C$ or $3°$ above absolute zero. The source was everywhere, pervading the whole universe: they had discovered the cool afterglow of the Big Bang.

Hubble had shown how fast the universe is expanding, and so we can imagine running the film backwards in time to see what it was like in the past. This reveals that some 14 billion years ago, all the material that we see today in the galaxies of stars would have been compacted together in a volume smaller than an atom. It is the explosion outwards that we refer to as the Big Bang.

Penzias and Wilson's measurement of the temperature of our large, mature cold universe gives a scale from which the heat of the early dense universe can be established. Their thermometer shows that the early moments of the universe would have been incredibly hot, trillions of degrees and more, far hotter than in the stars today. At such temperatures, conditions are similar to the localised heat produced when beams

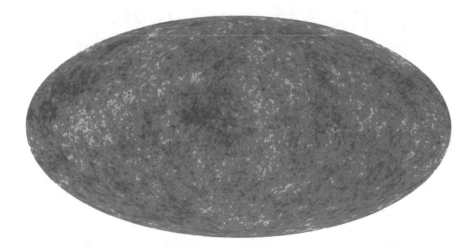

FIGURE 1.1 The microwave background as measured by 3 years of data from the WMAP satellite. The different shadings show temperature fluctuations of as little as two ten-thousandths of a degree, which are 13.7 billion years old and correspond to the seeds that grew to become the galaxies. (See also chapter 14.) (Source: NASA/WMAP Team.)

of particles such as electrons and protons smash into one another at high-energy particle accelerators. It is this link that enables high-energy particle physics to teach us about the formation and development of the very early universe.

These discoveries played a major part in establishing the Big Bang theory of genesis. Recently, experiments using special detectors in space have discovered small irregularities in the temperature of the background radiation as one scans across the sky (Figure 1.1). These subtle variations correspond to regions that are slightly hotter or cooler than the average, by amounts that are a trifling one thousandth of a degree, or even less. Their nature and distribution across the sky suggests that they may herald the birth of the first galaxies of stars less than a few hundred thousand years after the Big Bang.

So today we have a fairly broad picture of how the universe developed. As I write, we are at the threshold of a huge experiment that will reproduce the aftermath of the Big Bang in miniature. It is taking place at CERN in Geneva, at the Large Hadron Collider, known as the LHC, and will show what the universe was like when it was less than a hundredth of a billionth of a second old. My purpose in this book is to show how we have come to this understanding of the nature of the universe, of what matter is and where it came from. It is a heroic story that began thousands of years ago and whose greatest and most exciting consequences are soon to appear. By the end of this book, I will have set the scene for why the LHC is being built and speculate on what we might find there.

INWARD BOUND

Nature has buried its secrets deep, but not entirely hidden them. Clues to the restless agitation within its atomic architecture are everywhere: the static electricity released when brushing dry hair, sparks in the air and lightning, the radioactivity of natural rocks; the Aurorae Borealis caused by cosmic particles hurtling towards the north magnetic pole and the ability of a small compass needle to sense that magnetism thousands of miles away are just a few examples.

By ingeniously pursuing such clues, scientists have worked out how matter is made. Its basic seeds were formed in the searing cauldron of a Hot Big Bang, then fused into elements inside stars, leading to today's planets, Earth and life as the cool end-products of that creation.

The idea that there is a basic simplicity underlying the infinite variety surrounding us, originated in ancient Greece. In that philosophy, Earth, Fire, Air, and Water were the basic elements from which everything is made. Their idea was basically correct; it was the details that were wrong. By the 19th century, chemists had discovered nearly 90 atomic elements, which combine to make all substances, including the Greeks' classical four. Their Water, in pure form, requires two elements: hydrogen and oxygen. Air is dominantly made from nitrogen and oxygen with some carbon and traces of others. Fire is light emitted when hot atoms of carbon and oxygen are disrupted in the process of being burned into carbon dioxide. The Earth's crust contains most of the 90 naturally occurring elements, primarily oxygen, silicon, and iron, mixed with carbon, phosphorous, and many others that you may never have heard of, such as ruthenium, holmium, and rhodium.

An atom is the smallest piece of a chemical element that retains its elemental identity. With every breath, you inhale a million billion billion atoms of oxygen, which gives some idea of how small each one is. The seeds of the atoms that make everything on Earth today were cooked in a star some 5 billion years ago. So you are made of stuff that is as old as the planet, one third the age of the universe, but this is the first time that these atoms have gathered together in such a way that they think that they are you.

Atoms are not the smallest things. Whereas little more than a century ago atoms were thought to be small impenetrable objects, today we know that each has a rich labyrinth of inner structure where electrons whirl around a massive compact central nucleus.

Electrons are held in place, remote from the nucleus, by the electrical attraction of opposite charges, electrons being negatively charged and the atomic nucleus positively charged. A temperature of a few thousand degrees is sufficient to break this attraction completely and liberate all the electrons from within atoms. Even room temperature can be enough to release one or two; the ease with which electrons can be moved from one atom to another is the source of chemistry, biology, and life. Restricted to relatively cool conditions, the 19th century scientist was only aware of chemical activity; the heart of the atom — the nucleus — was hidden from view.

You are seeing these words because light is shining on the page and then being transmitted to your eyes; this illustrates the general idea, which will be common to the way that we study the make-up of matter, that there is a source of radiation (the light), the object under investigation (the page), and a detector (your eye). Inside that full stop are millions of carbon atoms and you will never be able to see the individual atoms, however closely you look with a magnifying glass. They are smaller than the wavelength of "visible" light and so cannot be resolved under an ordinary microscope.

Light is a form of electromagnetic radiation. Our eyes respond only to a very small part of the whole electromagnetic spectrum; but the whole of the spectrum is alive. Visible light is the strongest radiation given out by the Earth's nearest star, the Sun; and humans have evolved with eyes that register only this particular range. The whole spread of the electromagnetic spectrum is there. I can illustrate this by making an analogy with sound.

Imagine a piano keyboard with a rainbow painted on one octave in the middle. In the case of sound, you can hear a whole range of octaves but light — the rainbow that our eyes can see — is only a single octave in the electromagnetic piano. As you go from red light to blue, the wavelength halves; the wavelength of blue light is half that of red. The electromagnetic spectrum extends further in each direction. Beyond the blue horizon — where we find ultra-violet, x-rays, and gamma rays — the wavelength is smaller than in the visible rainbow; by contrast, at longer wavelengths and in the opposite direction, beyond the red, we have infra-red, microwaves and radio waves.

Ancient cave dwellers needed to see the dangers from wild animals; they had no need to develop eyes that could see radio stars. It is only since 1945 that we have opened up our vision across the fuller electromagnetic spectrum. The visions that have ensued have been no less dramatic than a Beethoven symphony would have been to a 13th century monk whose experiences ended with Gregorian chants restricted to the single octave of the bass clef.

We can sense the electromagnetic spectrum beyond the rainbow; our eyes cannot see infra-red radiation but our skin can feel this as heat. Modern heat-sensitive or infra-red cameras can "see" prowlers in the dark by the heat they give off. Bees and some insects can see into the ultra-violet, beyond the blue horizon of our vision, which gives them advantages in the Darwinian competition for survival. It is human ingenuity that has made machines that can extend our vision across the entire electromagnetic range, with radio telescopes on the ground and x-ray and infra-red telescopes in satellites complementing optical telescopes such as the Hubble space telescope. We have discovered marvels such as pulsars, quasars, and neutron stars, and have flirted with the environs of black holes. The visions beyond the rainbow have revealed deep truths about atoms and more.

Our inability to see atoms has to do with the fact that light acts like a wave, and waves do not scatter easily from small objects. To see things, the wavelength of the beam must be smaller than the thing you are looking at, and therefore to see molecules or atoms, you need illuminations whose wavelengths are similar to or smaller than them. To have any chance of seeing them, we have to go far beyond the blue horizon to wavelengths in the x-ray region and beyond.

X-rays are light with such small wavelength that they can be scattered by regular structures on the molecular scale, such as are found in crystals. The wavelength of x-rays is larger than the size of atoms, so these remain invisible. However, the distance between adjacent planes in the regular matrix within crystals is similar to the x-ray wavelength and so x-rays begin to discern the relative positions of things within crystals. This is known as "x-ray crystallography," which has become the main and revolutionary use of x-rays in science. Its most famous application was in the discovery of the structure of DNA, so I propose that modern genetics be promoted as applied physics.

It costs energy to produce radiation with short wavelengths. According to quantum theory, light and all radiation comes in staccato bundles, which act like particles, known as photons. The wavelength of the radiation and the energy of one of its photons are related. This is true not just for photons, but also for all particles. So here we see for the first time the rationale behind "high-energy" physics. Electric fields can accelerate electrically charged particles such as electrons up to energies limited only by technology and the cost to the taxpayer. The most powerful accelerators today can radiate light with the ability to resolve distances some hundred times smaller than the size of an atomic nucleus. Under these conditions, we can reveal the quarks, the constituent pieces from which nuclear particles such as protons and neutrons are made. If there are structures on distance scales smaller than these, it will require yet higher energies to reveal them. Experiments are being planned that will soon be able to resolve distances down to about 10^{-19} m, as tiny compared to a proton as is that proton to an entire atom of hydrogen.

THE FLY IN THE CATHEDRAL

If an atom were enlarged to the size of a cathedral, its nucleus would be no bigger than a fly. The accidental discovery of natural radioactivity by Becquerel in 1896 provided the tools with which the atom could be smashed and the nucleus revealed.

Metaphorically peel away the layers of the cosmic onion, deep into the heart of the atom and there, we believe, we will find the truly elementary particles from which matter is made. To the best measurements we can make, electrons appear to be fundamental. An atomic nucleus, by contrast, is not. Atomic nuclei are clusters of particles known as neutrons and protons, which in turn are made from more basic particles called quarks.

These are the most fundamental particles of matter that we know of. The electron and the quarks are like the letters of Nature's alphabet, the basic pieces from which all can be constructed; if there is something more basic, like the dot and dash of Morse code, we do not know for certain what it is. There is speculation that if you could magnify an electron or a quark another billion billion times, you would discover the underlying Morse code to be like strings, vibrating in a universe with more dimensions than our familiar three-dimensional space and single arrow of time. Whether this is science fiction or fact, and whether there is any deeper layer in the cosmic onion, are beyond our ability to answer at present. In this book I will limit to what is established and to what is likely to become so within the next decade.

Within the heart of the atomic nucleus, powerful forces are at work (see Figure 1.2). The fundamental quarks are gripped so tightly to each other that they occur only in groups, never in isolation: it is compact bundles of quarks that form what we call protons and neutrons. The force that in turn clusters the protons and neutrons together and builds the nucleus is so strong that vast energies can be released when its grip is broken: disrupt the effects of these strong forces and you can release nuclear power. By contrast, the heat from chemical reactions such as the burning of coal in a fire, and which involve only the electrons in the outer reaches of the atom, is trifling compared to the nuclear heat coming from the Sun or from the stars. The energy output from stars is so huge that they are visible in the night sky as we look back across space and

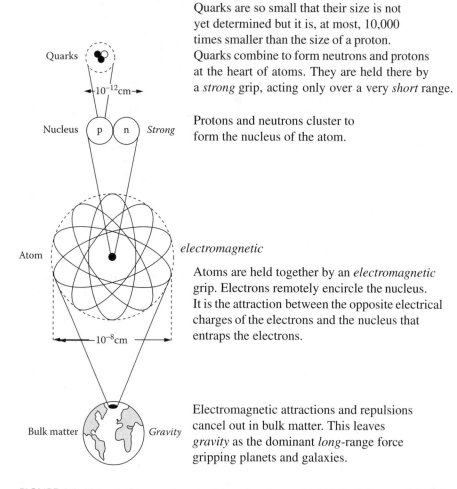

Quarks are so small that their size is not yet determined but it is, at most, 10,000 times smaller than the size of a proton. Quarks combine to form neutrons and protons at the heart of atoms. They are held there by a *strong* grip, acting only over a very *short* range.

Protons and neutrons cluster to form the nucleus of the atom.

electromagnetic

Atoms are held together by an *electromagnetic* grip. Electrons remotely encircle the nucleus. It is the attraction between the opposite electrical charges of the electrons and the nucleus that entraps the electrons.

Electromagnetic attractions and repulsions cancel out in bulk matter. This leaves *gravity* as the dominant *long*-range force gripping planets and galaxies.

FIGURE 1.2 Nature's forces act on particles of matter and build the bulk material of the universe.

time, in some cases receiving their light millions of years after they set out on their journey.

At the root of many of these nuclear processes is the transformation of neutrons into protons, which converts one element into another. Hydrogen and helium are fused together in stars, which can build them up into the more complex nuclei of heavier elements such as carbon, oxygen, iron, and the other important ingredients from which the Earth, the air and our bodies are formed. Although it has dramatic effects, the force that is responsible for this alchemy is rather feeble when acting in this way. It is over a thousand times weaker than the electromagnetic force and nearly a million times weaker than the strong nuclear force. However, in recent years it has become apparent that this so-called "weak" force is really a subtle manifestation of the more familiar electromagnetic force. This unity is hidden to our normal senses and even obscure when causing natural radioactivity of rocks, but becomes apparent in phenomena at extremely high energies, such as were prevalent very early in the life of the universe and are now accessible in experiments at particle accelerators.

These three distinct forces — the electromagnetic force, the strong force, and the weak force — together with gravity, control the behaviour of all bulk matter and of biological, chemical, and nuclear phenomena. The vast differences in their strengths are crucial to our existence. The reason for this disparity has been a long-standing puzzle, but one that may be about to be solved.

Another puzzling feature of the forces is the discriminatory manner in which they act. Gravity is exceptional in that it acts attractively between all particles. Electromagnetic interactions act only on particles carrying electrical charge, and the strong interaction in normal matter operates only within the atomic nucleus: electrons are remote from the nucleus, not least because they are unaffected by the strong force. Ghostly particles called neutrinos are only affected noticeably by the weak force and are, in consequence, not trapped in atoms. Thus, although neutrinos are crucial to our universe's existence, they are irrelevant in everyday chemistry.

The existence of atoms, and of gravitational and electromagnetic forces, defined the frontiers of the 19th century scientists' knowledge of the structure of matter and the fundamental forces. Natural radioactive decays led to the discovery of the atomic nucleus, and we became aware of the strong and weak nuclear forces during the first half of the 20th century. In the 1960s, huge machines were built, several kilometres in length, which could accelerate electrons or protons to almost the speed of light. These subatomic 'bullets' then smashed into targets of nuclear material, ploughing deep into the neutrons and protons within. For a fraction of a second, they were heated to temperatures higher than in any star. These experiments showed that neutrons and protons consist of more basic particles: the quarks. In addition, they revealed that the strong nuclear force was a remnant of a much more powerful force that clusters quarks together to build those neutrons and protons.

Under these very hot conditions, nuclear processes took on different aspects from those exhibited at the lower temperatures to which science had previously been limited. The strong force acting on quarks, and the electromagnetic force acting on electrons, began to show similarities to one another. Electromagnetism and radioactivity also seemed to be two different manifestations of a single force. Not only did the forces appear to be united at these extremes but, in addition, the fundamental quarks, electrons, and neutrinos started to appear as siblings rather than unrelated varieties of elementary particles. The asymmetry and disparity of our familiar cold world seem to be frozen remnants of symmetry and unity prevalent at ultra-high energies.

In the last quarter of the 20th century, a profound insight emerged: the material universe of today has emerged from a hot Big Bang, and the collisions between subatomic particles are capable of recreating momentarily the conditions that were present at that early epoch. This discovery had a dramatic impact on our understanding of the universe. The disciplines of high-energy physics (the study of subatomic phenomena) and cosmology or astrophysics had seemed far removed — the studies of the ultra-small and the far reaches of outer space. All this has now changed, and the point of union is the physics of the Big Bang.

Cosmologists now agree that our material universe emerged from the hot Big Bang — a compact fireball where extreme conditions of energy abounded. Discovering the nature of the atom a hundred years ago was relatively simple: atoms are ubiquitous in matter all around and teasing out their secrets could be done with apparatus on a table top. Investigating how matter emerged from Creation is another challenge entirely. There is no Big Bang apparatus for purchase in the scientific catalogues. The basic pieces that create the beams of particles, speed them to within an iota of the speed of light, smash them together and then record the results for analysis, all have to be made by teams of specialists. That we can do so is the culmination of a century of discovery and technological progress. It is a big and expensive endeavour, but it is the only way that we know how to answer such profound questions.

Collisions of high-energy particles create in a small region of space conditions that are feeble copies of that first Big Bang. But the highest manmade energies are far removed from the intensities that occurred then. The early universe has been likened to the ultimate high-energy physics laboratory and, it is argued, an elegance and beauty which existed then became obscured as the universe cooled. Our high-energy physics experiments have revealed a glimpse of that early unity.

Tempted by that vision, physicists seek the Grand Unified Theory of matter and forces: from their birth in the Big Bang to the structured universe that we now inhabit. The belief is that the strong, weak, and electromagnetic forces were originally one, and developed their separate identities as the universe expanded and cooled. Our normal experiences here in the coolness of Earth give poor insight into the processes at work in the Big Bang. The emerging Grand Unified Theories suggest the existence of new forms of particles, with exotic properties known as supersymmetry, which mirror those known to date.

These theories could be the breakthrough that will enable us to understand genesis as never before. But to make this science and not metaphysics, we must test the theories. One way is to push experiments to the highest energies possible and see ever-clearer visions of the Big Bang extremes. This will not be possible much longer, barring technological breakthroughs, because of the cost of building ever-more powerful machines. However, there is still much that can be done in the next 10 to 20 years that should reveal the successes or inadequacies of the Grand Unified Theories. The complementary approach is to look for exotic rare phenomena under cold conditions. Processes common in the Big Bang are not entirely absent in cold conditions — they are merely slowed down and thus rarer in occurrence. Thus, like archaeologists, some intrepid scientists seek evidence of rare events, relics of the Big Bang that by chance might be detected among the naturally occurring cosmic rays. There are also exciting new ventures using neutrinos as a way of looking into the heart of the Sun and giving the promise of "neutrino astronomy" as a growth science in the next decades. However, the main theme of this book will be to tell how we came to know of the basic seeds of matter and the forces that control them.

In a nutshell, this is how matter as we know it today evolved from the Big Bang. Out of that fireball, the quarks, electrons, and neutrinos were the survivors while the universe was still very young and hot. As it cooled, the quarks were gripped to one another, forming protons and neutrons. The mutual gravitational attraction among these gathered them into large clouds that were primaeval stars. As they bumped into one another in the heart of stars, the protons and neutrons built up the seeds of heavier elements. Some stars become unstable and explode, ejecting these atomic nuclei into space where they trap electrons to form atoms of matter in the forms familiar today. We believe that the explosion of such a "supernova" occurred some 5 billion years ago when our solar system was forming. It was debris from that long-dead supernova that made you and me.

By performing special experiments, we can, in effect, reverse the process and observe matter change back into its original primaeval forms. Heat matter to a few thousand degrees and its atoms ionize — electrons are separated from the central nuclei. That is how it is inside the Sun. The Sun is a plasma — gases of electrically charged electrons and protons swirling independently. At even higher temperatures, typical of the conditions that can be reached in relatively small high-energy accelerators, the nuclei are disrupted into their constituent protons and neutrons. At yet higher energies, these in turn "melt" into a plasma of freely flowing quarks. (See Figures 1.3 and 1.4 and accompanying text.)

The purpose of this book is to tell you how we came to know of the electron and the quarks, who they are, how they behave, and what questions confront us. In the early chapters we meet the history and discoveries that led to the identification of microscopic matter, of its basic building blocks, and the forces that cluster them to build the large-scale universe in which we live. The later chapters describe the new insights that have caused us to recognise the unity in nature under extreme conditions such as prevailed in the early universe. We conclude with a brief description of the current theories, their consequences and tests, and their impact on cosmology.

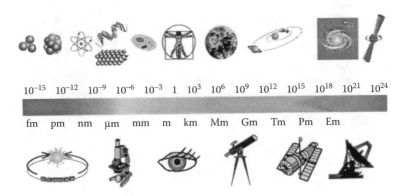

FIGURE 1.3 Comparisons with the human scale and beyond normal vision. In the small scale, 10^{-6} m is known as 1 micron, 10^{-9} m is 1 nanometre, and 10^{-15} m is 1 fermi. (Source: CERN.)

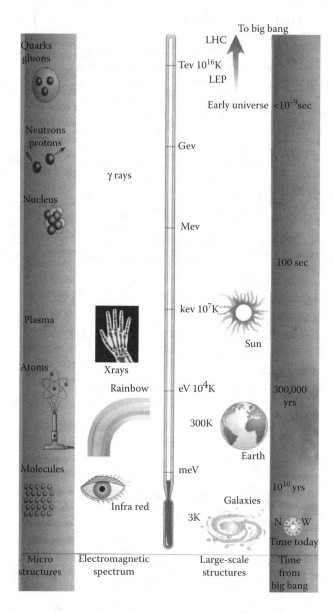

FIGURE 1.4 The correspondence between scales of temperature and energy in electron-volts (eV). (Source: *Particle Physics – A Very Short Introduction*, Frank Close.)

The changing face of matter as temperature increases.

Energy is profoundly linked to temperature. If you have a vast number of particles bumping into one another, transferring energy from one to the next so that the whole is at some fixed temperature, the average energy of the individual particles can be expressed in eV, keV, etc. Room temperature corresponds to about 1/40 eV or 0.025 eV. Perhaps easier will be to use the measure of 1 eV $\rightleftharpoons$ 10^4 K (where K refers to Kelvin, the absolute measure of temperature; absolute zero 0 K = −273 Celsius, and room temperature is about 300 K).

Fire a rocket upwards with enough energy and it can escape the gravitational pull of the Earth; give an electron in an atom enough energy and it can escape the electrical pull of the atomic nucleus. In many molecules, the electrons will be liberated by an energy of fractions of an electron-volt; so room temperature can be sufficient to do this, which is the source of chemistry, biology, and life. Atoms of hydrogen will survive at energies below 1 eV, which in temperature terms is on the order of 10^4 K. Such temperatures do not occur normally on Earth (other than specific examples such as some industrial furnaces, carbon arc lights, and scientific apparatus) and so atoms are the norm here. However, in the centre of the Sun, the temperature is some 10^7 K, or in energy terms 1 keV; atoms cannot survive such conditions.

At temperatures above 10^{10} K, there is enough energy available that it can be converted into particles, such as electrons. An individual electron has a mass of 0.5 MeV/c^2, and so it requires 0.5 MeV of energy to "congeal" into an electron. As we shall see later, this cannot happen spontaneously; an electron and its antimatter counterpart — the positron — must be created as a pair. So, 1 MeV energy is needed for "electron positron creation" to occur. Analogously, 2 GeV energy is needed to create a proton and its antiproton. Such energies are easy to generate in nuclear laboratories and particle accelerators today; they were the norm in the very early universe, and it was in those first moments that the basic particles of matter (and antimatter) were formed. They are far beyond normal experience, so matter on Earth survives and light does not spontaneously convert into matter and antimatter before our eyes.

Energy Accounts: from macro to micro

In macroscopic physics we keep our energy accounts in Joules, or in large scale industries, mega, or terajoules. In atomic, nuclear, and particle physics, the energies involved are trifling in comparison. If an electron, which is electrically charged, is accelerated by the electric field of a 1-V battery, it will gain an energy of 1.6×10^{-19} J. Even when rushing at near the speed of light, as in accelerators such as at CERN in Geneva, the energy still only reaches the order of 10^{-8} J, one hundredth of a millionth of a Joule. Such small numbers get messy and so it is traditional to use a different measure, known as the electron-volt or eV. We said above that when accelerated by the electric field of a 1-V battery, it will gain an energy of 1.6×10^{-19} J and it is this that we define as 1 eV.

Now the energies involved in subatomic physics become manageable. We call 10^3 eV a kilo-eV or keV; a million (mega), 10^6 eV is 1 MeV; a billion (giga), 10^9 eV is 1 GeV, and the latest experiments are entering the "tera" or 10^{12} eV, 1 TeV, region.

Einstein's famous equation $E = mc^2$ tells us that energy can be exchanged for mass, and vice versa, the "exchange rate" being c^2, the square of the velocity of light. The electron has a mass of 9×10^{-31} kg. Once again, such numbers are messy and so we use $E = mc^2$ to quantify mass and energy, which gives about 0.5 MeV for the energy of a single electron at rest; we traditionally state its mass as 0.5 MeV/c^2. The mass of a proton in these units is 938 MeV/c^2, which is nearly 1 GeV/c^2.

2 Atoms

As we look at the world around us there is nothing immediately that says everything is made of atoms. An atom is the smallest piece of a chemical element and consists of a tiny massive central core surrounded by electrons. It is the gyration of the electrons, switching ephemerally between neighbouring atoms that link one to another building up molecules and bulk matter. Richard Feynman, one of the greatest physicists of the 20th century, when asked which piece of knowledge he would most wish to be passed on to survivors of some future global catastrophe, said: "We're made of atoms."

This is one of the most profound and far-reaching discoveries. The story of how we arrived at this insight begins over 4500 years ago in Greece, where Thales realized that any substance could be classified as solid, liquid, or gas. This might seem rather obvious to us today as we look back with several millennia of accumulated wisdom; but at that time, when many of our modern "enlightened" nations were still in the dark ages, it was an insight of genius. In addition, he went further: water can exist in each of these forms — so might it be the case that all matter is nothing more than water?

The article of faith that a basic simplicity lurks behind the scenes of Nature's infinite variety waxed and waned over the centuries. It was Thales' followers who extended his idea by postulating that matter is made from the four elements — earth, fire, air, and water — and Democritus who in 585 BC suggested that everything ultimately is built from small indivisible particles — in Greek $\alpha\tau o\mu o\sigma$ — atoms.

According to folklore it was the smell of baking that inspired the idea that small particles of bread existed beyond vision. The cycle of weather reinforced this: a puddle of water on the ground gradually dries out, disappears, and then falls later as rain. Water was one of their elements; they reasoned that there must be particles of water that evaporate, coalesce in clouds, and fall to Earth, so that the water is conserved even though the little particles were too small to see. Combining the ideas of Thales and Democritus would have given the seeds of modern scientific thought, namely that the world is built from a few basic indivisible constituents. However, their ideas lay dormant for over 2000 years.

In 1808, John Dalton, an English chemist, proposed the modern atomic theory of matter. Chemistry was becoming a quantitative science and had shown that a wide variety of substances could be formed by combining different quantities of a few elements such as hydrogen, carbon, oxygen, sodium, chlorine, and so forth. Over 90 of these elementary substances are now known.

Dalton realised that if each element is made from atoms — very tiny objects, which join together to build up the substances that are large enough to see — he could explain chemistry's findings. Combining atoms of various elements together made molecules of non-elementary substances. Furthermore, he believed that atoms were indivisible; indeed, it was for this reason that he honoured Democritus and named them 'atoms.'

The modern atomic elements include gases at room temperature, such as hydrogen and oxygen; liquid mercury; and solid gold. The various elements have different values (do you prefer gold or carbon?) and can occur in different forms (carbon may be graphite or compressed into diamond — so choose your answer to the previous question carefully). Silicon is uniquely suitable for modern electronics. The malleability and toughness of lead, the conductivity of copper, and the lustre of gold are but some of the specific attributes that give each a special character. In combinations, they form molecules; from some hundred atomic elements there are countless molecular combinations that form the substances of the universe.

We can smell molecules but normally cannot see individual ones. A simple demonstration can give an idea of their size. First, fill a tray with water and then sprinkle some flour (or better, if you can get some, lycapodium powder) so as to make the surface easy to see. Next, take a small drop of oil and very gently release it onto the surface of the water. It spreads and eventually stops. The reason is that the molecules of oil that were originally all clustered in the spherical droplet have tumbled over each other until they are just one molecule thick. At that point they can go no further; they have reached their natural extent.

You can see the area they have filled; and if you know the amount that you started with, you can compare that volume (a three-dimensional quantity) with the area (two-dimensional) to determine the extent of the one dimension missing. The area divided into the volume thus gives the height of the molecules, which is revealed to be about a hundred thousandth of a millimetre. About 10,000 such molecules could be spread across the diameter of a human hair. So individual molecules are beyond normal vision, but not so far away.

Oil molecules are quite long, with tens of atoms loosely clustered together. If individual atoms were laid end to end, up to a million would cover the thickness of a hair. Visualise the crowd at a major football game; multiply it ten times; now imagine that number in the width of a hair. Atoms are probably at the limits of our imagination.

By the beginning of the 20th century it was becoming increasingly obvious that atoms are not the most fundamental entities in nature. The sequence of events that led to the unravelling of the inner structure of atoms is well known, but for future reference some features are worth mentioning here.

The first indirect hints emerged around 1869 when Dmitri Mendeleev discovered that listing the atomic elements from the lightest — hydrogen — up to the heaviest then known — uranium — caused elements with similar properties to recur at regular intervals (Table 2.1). If each element were truly independent of all the others, then any similarities among them would have been merely coincidental and occurred randomly. The empirical regularity was striking, and today is understood as arising from the fact that atoms are not elementary, but are instead complex systems built from common constituents: electrons surrounding a compact nucleus. The whole is held together by the electromagnetic attraction of opposite charges — electrons being negatively, and the nucleus positively, charged. The experiments that led to this picture of atomic structure were made by Rutherford and co-workers little more than 90 years ago and were the fruits of a sequence of discoveries spanning several years.

Deep within matter, atoms are announcing their presence by spitting out rays into the air. For billions of years they have done so, forming the elements of life,

Table 2.1a Mendeleev's Periodic Table of the Atomic Elements

Typische elemente							
	Li = 7 / Na = 23 group						
II = 1			K = 39	Rb = 85	Ca = 133	—	—
			Ca = 40	Sr = 87	Ba = 137	—	—
			*	?Yt = 88?	?Di = 138?	Er = 178?	Th = 231
			Ti = 48?	Zr = 90	Ce = 140?	?La = 180?	—
			V = 51	Nb = 94	—	Ta = 182	U = 240
			Cr = 52	Mo = 96	—	W = 184	—
			Mn = 55	Ru = 104	—	—	—
			Fe = 56	Rh = 104	—	Qa = 195?	—
			Co = 59	Pd = 106	—	Ir = 197	—
	Li = 7	Na = 23	Ni = 59	Ag = 108	—	Pt = 198?	—
	Be = 9.4	Mg = 24	Cu = 63	Cd = 112	—	Au = 199?	—
	B = 11	Al = 27.3	Zn = 65	In = 113	—	Hg = 200	—
	C = 12	Si = 28	*	Sn = 118	—	Ti = 204	—
	N = 14	P = 31	As = 75	Sb = 122	—	Pb = 207	—
	O = .16	S = 32	Se = 78	Te = 125?	—	Bi = 208	—
	F = 19	Cl = 35.5	Br = 80	J = 127	—	—	—

The periodic table of the atomic elements, showing the elements known to Mendeleev in 1869 and its modern version. Scandium, gallium, and germanium were successfully predicted by Mendeleev to fill the sites denoted by *.

Table 2.1b The Modern Periodic Table of the Elements

Key

relative atomic mass
atomic symbol
name
atomic (proton) number

Example: 1.0 **H** hydrogen 1

(1)	(2)	(3)	(4)	(5)	(6)	(7)	(8)	(9)	(10)	(11)	(12)	(13)	(14)	(15)	(16)	(17)	(18)
6.9 **Li** lithium 3	9.0 **Be** beryllium 4											10.8 **B** boron 5	12.0 **C** carbon 6	14.0 **N** nitrogen 7	16.0 **O** oxygen 8	19.0 **F** fluorine 9	20.2 **Ne** neon 10
23.0 **Na** sodium 11	24.3 **Mg** magnesium 12											27.0 **Al** aluminium 13	28.1 **Si** silicon 14	31.0 **P** phosphorus 15	32.1 **S** sulfur 16	35.5 **Cl** chlorine 17	39.9 **Ar** argon 18
39.1 **K** potassium 19	40.1 **Ca** calcium 20	45.0 **Sc** scandium 21	47.9 **Ti** titanium 22	50.9 **V** vanadium 23	52.0 **Cr** chromium 24	54.9 **Mn** manganese 25	55.8 **Fe** iron 26	58.9 **Co** cobalt 27	58.7 **Ni** nickel 28	63.5 **Cu** copper 29	65.4 **Zn** zinc 30	69.7 **Ga** gallium 31	72.6 **Ge** germanium 32	74.9 **As** arsenic 33	79.0 **Se** selenium 34	79.9 **Br** bromine 35	83.8 **Kr** krypton 36
85.5 **Rb** rubidium 37	87.6 **Sr** strontium 38	88.9 **Y** yttrium 39	91.2 **Zr** zirconium 40	92.9 **Nb** niobium 41	95.9 **Mo** molybdenum 42	[98] **Tc** technetium 43	101.1 **Ru** ruthenium 44	102.9 **Rh** rhodium 45	106.4 **Pd** palladium 46	107.9 **Ag** silver 47	112.4 **Cd** cadmium 48	114.8 **In** indium 49	118.7 **Sn** tin 50	121.8 **Sb** antimony 51	127.6 **Te** tellurium 52	126.9 **I** iodine 53	131.3 **Xe** xenon 54
132.9 **Cs** caesium 55	137.3 **Ba** barium 56	138.9 **La*** lanthanum 57	178.5 **Hf** hafnium 72	180.9 **Ta** tantalum 73	183.8 **W** tungsten 74	186.2 **Re** rhenium 75	190.2 **Os** osmium 76	192.2 **Ir** iridium 77	195.1 **Pt** platinum 78	197.0 **Au** gold 79	200.6 **Hg** mercury 80	204.4 **Tl** thallium 81	207.2 **Pb** lead 82	209.0 **Bi** bismuth 83	[209] **Po** polonium 84	[210] **At** astatine 85	[222] **Rn** radon 86
[223] **Fr** francium 87	[226] **Ra** radium 88	[227] **Ac*** actinium 89	[261] **Rf** rutherfordium 104	[262] **Db** dubnium 105	[266] **Sg** seaborgium 106	[264] **Bh** bohrium 107	[277] **Hs** hassium 108	[268] **Mt** meitnerium 109	[271] **Ds** darmstadtium 110	[272] **Rg** roentgenium 111							

Elements with atomic numbers 112-116 have been reported but not fully authenticated

The Lanthanides (atomic numbers 58-71) and the Actinides(atomic numbers 90-103) have been omitted.

maintaining the inner heat of the Earth, awaiting discovery by science. If a single moment marks the start of modern science, it was the Friday evening of 8th November 1895 when Röntgen discovered x-rays. They are most familiar today by their ability to cast shadows of broken bones. It was Röntgen's shadow graph of the bones of his wife's hand, with her wedding ring clearly visible, that has become one of the most famous images in photographic history and which within weeks turned him into an international celebrity. The medical implications were immediately realised and the first images of bone fractures were being made by January 1896.

That x-rays existed with remarkable properties was immediately apparent, but what they are and how they are produced was a mystery. Today we know that they are a form of electromagnetic radiation — essentially light whose wavelengths are much shorter than those that our eyes record as colours.

Jiggle a stick from side to side on the surface of a still pond, and a wave will spread out. A cork floating some distance away will start wobbling when the wave reaches it. Energy has been transferred from the stick to the cork. This energy has been carried by the wave. When an electric charge is accelerated or jiggled, an "electromagnetic wave" travels outwards through space. A charge some distance away will be set in motion when the wave arrives. The wave has transported energy from the source to the receiver. A familiar example is an oscillating charge in a radio transmitter. This generates an electromagnetic wave, which transports energy to the charges in your radio aerial.

Electromagnetic waves and water waves differ in important ways however. The speed at which water waves travel depends on the distance between successive peaks and troughs — the wavelength. In contrast, all electromagnetic waves travel at the same speed — the speed of light. The only difference between light and radio waves is their wavelength — radio wavelengths are several metres, the visible light of the rainbow only about a hundred thousandth of a centimetre (hundreds of nanometres). The peaks and troughs of a short wave pass a point more frequently than do those of a long wave. It is these oscillations that transport the energy; thus, for two waves with the same amplitude, more energy is carried by the high-frequency (short wavelength) radiation than by the low frequency (long wavelength). Consequently, radio waves have low energy, visible light higher energy, and x-rays and γ-rays have in turn higher still. (See Table 2.2.)

That is now common knowledge, but at the end of the 19th century the pressing questions were to explain what x-rays are and how they are formed. In the course of sorting this out, a host of other discoveries were made, most notable among which was that atoms have a labyrinthine internal structure.

In Paris, Henri Becquerel had learned of Röntgen's x-rays in the first week of January 1896. By this time he was 44 years old, had a strong track record in research into phosphorescence, uranium compounds, and photography, and was a member of the French Academie des Sciences. The news of of Röntgen's discovery and the startling photographic images immediately resonated with Becquerel's experiences and made him wonder if x-rays and phosphorescence were related.

Once he had the idea, it was obvious how to test it. First he exposed some phosphorescent crystals to sunlight for several hours so they were energised as usual. Then he wrapped them in opaque paper, placed the package on top of photographic

Table 2.2 Electromagnetic Radiation

Jiggle a stick from side to side on the surface of a still pond, and a wave will spread out. A cork floating some distance away will start wobbling when the wave reaches it. Energy has been transferred from the stick to the cork. This energy has been carried by the wave.

If an electric charge is accelerated or jiggled, an 'electromagnetic wave' is transmitted through space. A charge some distance away will be set in motion when the wave arrives. The electromagnetic wave has transported energy from the source to the receiver. A familiar example is an oscillating charge in a radio transmitter. This generates an electromagnetic wave, which transports energy to the charges in your radio aerial.

The speed that water waves travel depends on the distance between successive peaks and troughs (the wavelength). In contrast all electromagnetic waves travel at the same speed–the speed of light. The only difference between light and radio waves is one of wavelength–radio wavelengths are several m, lit only about a hundred thousandth of a cm. In the quantum theory, the energy of electromagnetic radiation is inversely proportional to the wavelength. Consequently, quanta of radio waves have low energy, and those of light have intermediate energy, while x-rays have higher energy, and γ-rays have the highest energies.

The Electromagnetic Spectrum

Wavelength cm	Example		Use/manifestation
10^{-12} 10^{-11}	Photons in accelerators		Short wavelengths resolve minute structures in molecules and subatomic phenomena. Also used in biological research
10^{-10} 10^{-9}	γ-rays		
10^{-8} 10^{-7}	x-rays		x-ray photography. Penetrates skin. Stopped by hard tissue such as bone
10^{-6} 10^{-5}	Ultra violet		Sun tan
10^{-4}	Visible spectrum	Blue Green Yellow Orange Red	Eye
10^{-3} 10^{-2} 10^{-1}	Infra-red		Heat. Infrared photography.
1	Microwaves		Microwave cooking cosmic background
10	UHF		
100	VHF		
10^3	TV, FM radio		Radio and TV broadcasting
10^4	Short wave		
10^5 (1000 metre)	Medium wave (AM)		
1500 metres	BBC Radio 4, Long wave		

emulsion and put them all in a dark drawer. If the fluorescence emitted only visible light, none would get through the opaque paper to reach the photographic emulsion, whereas any x-rays emitted by the crystal would pass uninterrupted to the plates and fog them. As an extra test he placed some metal pieces between the package and the photographic material so that even x-rays would be blocked and leave a silhouette of the metal in the resulting image.

When he developed the plates he found that they had indeed been exposed and, most important, contained shadows of the metal plates. He noticed that uranium compounds seemed to produce strong images and hence inferred them to be particularly good sources of radiation. This was an important discovery, but he was wrong in his belief that exposure to sunlight had provided the energy that caused the process.

The true secret was still to be revealed. It was at the end of February 1896, as he continued his experiments, that he made his major discovery: uranium radiates energy spontaneously without need of prior stimulation such as sunlight. In fact, sunlight has nothing to do with it, as serendipity was about to reveal.

The final week of February in Paris was overcast. Mistakenly believeing that sunlight was needed to start the effect, he saw no point in continuing until the weather improved. As luck would have it, it remained cloudy all week and by 1 March he was becoming frustrated. At this point, for want of something to do, he decided to develop the plates anyway, expecting, as he later reported, to find a feeble image at best. To his surprise the images were as sharp as before, showing that the activiation took place in the dark.

Although this discovery of spontaneous 'radioactivity' is today recognised as seminal, it is ironic that at the time it made no special impact. The real birth of the radioactive era was when Pierre and Marie Curie discovered radioactivity in other elements, in particular radium, where the effect was so powerful that the radium glowed in the dark. (It was Marie Curie who invented the term 'radioactivity'). By 1903 the significance was fully realised and it was appropriate that the Curies shared the Nobel Prize with Becquerel.

Becquerel's discovery illustrates how science does not always proceed in ordered steps. The structure of the atom was not then known (indeed, it would be another 16 years before Rutherford disentangled its electron-orbiting-nucleus nature) and the source of Becquerel's radiation was still half a century away from being understood. The nuclei of the uranium atoms in his potassium-uranium-sulphate mixture were spontaneously breaking up, producing the radiations and a lot of energy — a property that eventually led to the development of the atom bomb and nuclear power.[1]

We are moving towards the modern picture of atoms, which have an inner structure held together by the mutual attraction of opposites, in the form of negative and positive electric charges. The negatively charged electron was discovered by J.J. Thomson in 1897. Electrons flow through wires as current and power industrial society; they travel through the labyrinths of our central nervous system and maintain our consciousness; and their motions from one atom to another underpin chemistry, biology, and life. The electron is the lightest particle with electric charge; it is stable and ubiquitous. Our limbs, bodily features, and indeed the shapes of all solid structures are dictated by the electrons gyrating at the periphery of atoms.

Gravity rules the universe, but it is electromagnetic forces, and their agents, the electrons, that give shape, form, and structure, especially here on Earth. The electron

---

[1] The story of uranium, and the chain of discovery leading to the atomic age, is well documented in Rhodes' *The Making of the Atomic Bomb* (see Suggestions for Further Reading).

is present in all space and for all time: modern theory suggests that the electron was the first material inhabitant of the universe at the act of Creation; and if the universe expands forever, electrons will probably be among its last remnants when the lights go out.

Take a deep breath. Each one of the multitude of oxygen atoms that you inhale contains negatively charged electrons. The fact that you have not become highly electrified is because each of those atoms also contains a corresponding amount of positive charge that balances and annuls the negative electrons. This simple demonstration makes the need for some positive charge within atoms almost obvious. The story of how the atom's structure was solved is less direct.

In October 1895, one month before Röntgen discovered x-rays and two months before the dramatic announcement of their discovery, a young New Zealander, Ernest Rutherford, had left home and travelled half way around the world to England and the Cavendish laboratory in Cambridge. When Rutherford first arrived in Cambridge in the autumn of 1895, Thomson was already deep into his research, his excitement intensifying when he learned that Röntgen's mystery rays ionized gases, liberating electric charge from within atoms, and that Becquerel had discovered radioactivity. A compelling new field was opening up, Thomson was near its frontier, and a brilliant young research student had arrived fresh in his laboratory. It was natural for Thomson to guide Rutherford into this new field.

Becquerel was continuing to improve his original measurement involving uranium; and while Marie Curie was trying to determine what elements other than uranium are sources of the rays, Rutherford set about using the rays as a tool, a convenient source for ionizing gases. This was but a brief interlude as he quickly realized that the greater question was the nature of the emanations themselves. He turned the focus of his enquiry on its head and used the ionization of gases as a means of studying radioactivity, rather than the other way around.

Rutherford, while working with Thomson at Cambridge between 1896 and 1900, showed that Becquerel's radiation contained three distinct components, which he named α (alpha), β (beta), and γ (gamma). The gamma radiation turned out to be electromagnetic radiation of extremely high energies, much higher than even x-rays. The beta radiation consisted of particles which were soon shown to be negatively charged electrons, and the alpha particles were massive, positively charged entities — known to be the nuclei of helium atoms (see Chapter 3). Having isolated these three components, though not yet knowing how they were formed nor what their true nature was, Rutherford exploited them to study atomic structure.

In 1902, Rutherford and Soddy discovered that some atoms could spontaneously disintegrate and produce other atoms. In the same decade, Pierre Curie and Marie Curie-Sklodowska discovered new radioactive elements — radium and polonium — in the products of uranium's disintegration. Suspicion grew that atoms had an inner structure which differed only slightly between one atom and the next. Small changes in this inner structure would convert one type of atom into another. Electrons were already well known from Thomson's work and are almost 2000 times lighter than the lightest atom (hydrogen): if electrons were one of the elementary ingredients that built up atoms, what else was there? The revelation of the inner atom, hardly dreamed of in 1895, was by the turn of the century the emerging challenge.

Electrons were seen to be deflected by magnetic fields and attracted towards positively charged objects. The rule 'like charges repel, unlike attract' showed that electrons carried negative electrical charge. When some of the negative charge in one atom is attracted by the positive charge of another, then the two atoms grip one another, binding together to form molecules and macroscopic matter. This insight begs the question of how atoms are constructed: how are the positive and negative charges arranged? What causes transmutation? What is the secret power of radioactivity?

To begin answering these questions, first it was necessary to identify precisely what carried the positive charge in atoms, and this needed some way of looking inside them. With brilliant directness, Rutherford went straight to the heart of the problem.

If you strike a sheet of lead with a hammer, you are hitting millions of its atoms at a time. The way that the sheet is bent by the blow will reveal how the atoms and molecules are bound together giving the metal its strength, but you learn nothing about individual atoms this way. To have a good chance of striking a single atom, you need to hit with something no bigger than it is. If you can do so, then you may learn something about the atoms themselves, in particular the way that the negative and positive electrical charges are distributed within them.

Ever since Rutherford had first isolated the alpha particle in 1899, he had been working to establish its identity. As a spin-off, he discovered that alpha particles are good atomic probes. First, alpha particles are emitted from atoms and so are much smaller than them. Second, they have a positive electrical charge and so will be repelled by the positive charges in the atoms.[2]

The alpha particles were moving faster than a speeding bullet, yet a thin sheet of mica deflected them slightly. Rutherford calculated that the electric forces within the mica must be immensely powerful compared with anything then known. Force fields of such strength in air would give sparks, and the only explanation he could think of was that these powerful electric fields must exist only within exceedingly small regions, smaller than even an atom. From this came his inspired guess: these intense electric fields are what hold the electrons in their atomic prisons and are capable of deflecting the swift alpha particles. Rutherford's genius once again came to the fore as he realised that he could now use the alpha particles to explore within atoms: from the way that beams of speeding alpha particles were deflected by atoms, he would be able to deduce the atomic electrical structure. Thus the scattering of alpha particles became one of the research problems on Rutherford's 1907 list as he started at Manchester.

Alpha particles are several thousand times heavier than electrons. Their relatively large bulk knocks lightweight electrons out of atoms, while their own motion continues almost undisturbed. Thomson, the discoverer of the electron, believed that positive charges were spread diffusely throughout an atom, possibly carried by light particles

---

[2] You might wonder why alpha particles were not suggested as the carriers of the positive charges in atoms. There were two main reasons. First, their mass was about four times that of hydrogen and so you could not build hydrogen that way. Second, their charge was twice that of the electron (and of opposite sign, of course), and it was felt most natural that the elementary units of positive and negative electric charges should be the same.

such as the electron. If his idea was correct, then the massive alpha particles would plough straight through the atoms.

Rutherford set about investigating this with his assistant Hans Geiger. They detected the alpha particles by the faint flashes that they gave upon hitting a 'scintilator:' a glass plate that had been coated with zinc sulphide. The pattern of deflections confirmed that there are intense electric fields in atoms, but their results were plagued by an incessant problem of stray scattered alphas that they could not explain. So Ernest Marsden, a student of Rutherford's, was given the task of seeing if any alpha particles might bounce back from thin wafers of gold.

One day in 1909, Marsden told Rutherford that most of them indeed passed straight through, but about one in 10,000 bounced back. Can cannon-balls recoil from peas? To be deflected through 90° after hitting a gold foil that is only 1/10,000 of a centimetre thick, as Marsden's results showed, required that they had been subjected to electric forces that were a thousand million times stronger than anything then known and far beyond anything that Rutherford and Geiger had been measuring. Rutherford later remarked that it was the most incredible event that had happened in his life: 'It was as if you had fired a 15-inch shell at a piece of tissue paper and it came back and hit you.' Somewhere in the gold atoms must be concentrations of material much more massive than alpha particles.

Rutherford spent a year puzzling about this and in 1911 announced his solution: all of the atom's positive charge and most of its mass are contained in a compact nucleus at the centre. The nucleus occupies less than 10^{-12} of the atomic volume — hence the rarity of the violent collisions — and the electrons are spread diffusely around outside. Rutherford computed how frequently alpha particles would be scattered through various angles and how much energy they would lose if a positively charged, dense nucleus was responsible for their recoil. During the next 2 years, Marsden and Hans Geiger scattered alpha particles from a variety of substances and confirmed Rutherford's theory of the nuclear atom.

Rutherford's original calculation left him astonished. In his notes (Figure 2.1) preserved in the library at Cambridge, his excitement is clear, for his handwriting became almost illegible as he wrote 'it is seen that the charged centre is very small compared with the radius of the atom.'

In this we have essentially the picture of the nuclear atom that has survived for the subsequent century. Negatively charged electrons and positive nuclei in perfect balance make a neutral atom. By contrast, in mass it is no contest as the positive nuclei outweigh the negative electrons by several thousands. Our matter is composed of 'Brobdignagian' positives and 'Lilliputian' negatives; negative and positive electrical charges balance neatly but in a very lopsided asymmetrical fashion.[3]

While this left no doubt that the positive charge is situated at the centre, there was still a puzzle about what the flighty electrons did. Rutherford had suggested that they orbited around the central nucleus so that an atom's structure is analogous to the solar system, the essential differences being the overall scale and that there is electromagnetic instead of gravitational attraction. This is an oft-quoted analogy, but

---

[3] This inbuilt asymmetry at the heart of matter is crucial to existence. For a detailed discussion see *Lucifer's Legacy — The Meaning of Asymmetry* by F.E. Close (Oxford University Press, 2000).

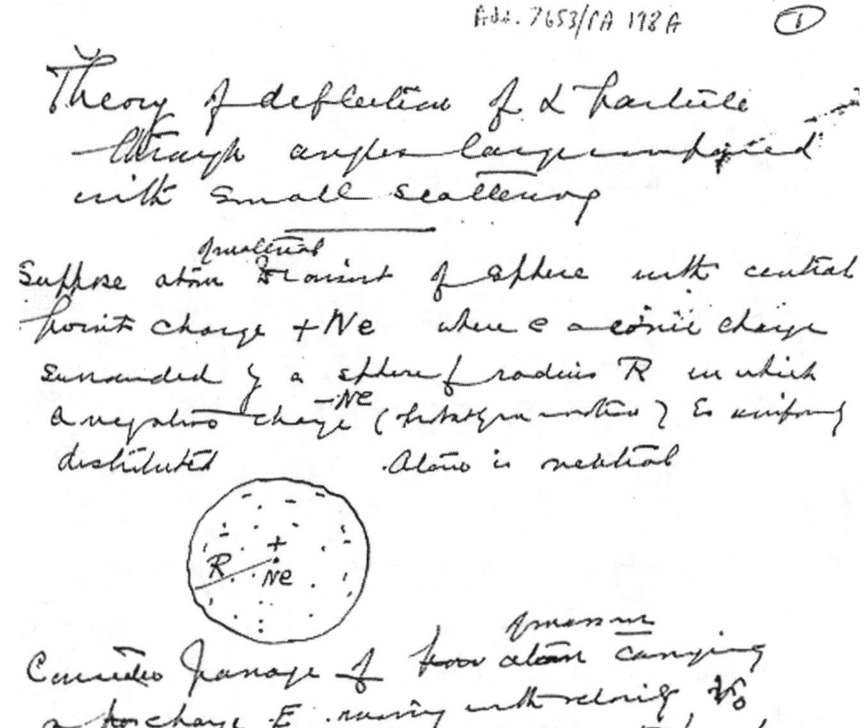

FIGURE 2.1 Rutherford's discovery of the atomic nucleus. These two pages of notes are where Rutherford first showed that the atomic nucleus is 'very small compared with the radius of the atom.' He has used simply the conservation of energy, equating the kinetic energy of the alpha particle when far from the nucleus to its electrostatic potential energy when it has momentarily come to rest at the point of closest approach. (With permission of Syndics of Cambridge University.) (*Continued.*)

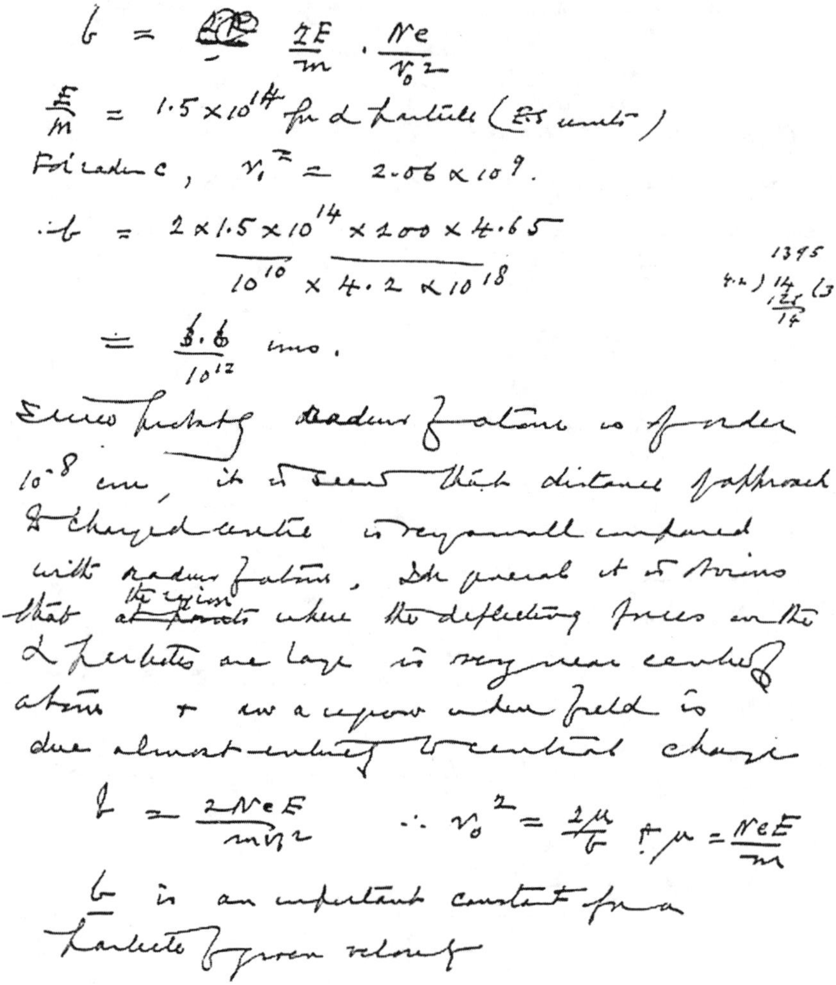

$$b = \frac{2E}{m} \cdot \frac{Ne}{v_0^2}$$

$$\frac{E}{m} = 1.5 \times 10^{14} \text{ for a particle (E.S. units)}$$

For radium c, $v_1^2 = 2.06 \times 10^9$.

$$\therefore b = \frac{2 \times 1.5 \times 10^{14} \times 200 \times 4.65}{10^{10} \times 4.2 \times 10^{18}}$$

$$= \frac{3.6}{10^{12}} \text{ cms.}$$

$$1395$$
$$4.2\,)\,\frac{14}{\frac{12}{14}}\,(3$$

Since probable radius of atom is of order 10^{-8} cm, it is seen that distance of approach to charged centre is very small compared with radius of atom. It is proved it is seen that at the region where the deflecting forces on the α particles are large is very near centre of atom + we a region where field is due almost entirely to central charge

$$b = \frac{2NeE}{mv_0^2} \qquad \therefore v_0^2 = \frac{2\mu}{b} \text{ if } \mu = \frac{NeE}{m}$$

b is an important constant for a particle of given velocity

FIGURE 2.1 (Continued).

a poor one for several reasons — one being that in reality the atom is far emptier than the solar system.

In the solar system, our distance from the Sun is 100 times larger than the diameter of the Sun itself; the atom is far emptier, with a factor of 10,000 as the corresponding ratio between the radius of an atom of hydrogen and the extent of its central nucleus. And this emptiness continues. Individual protons and neutrons are in turn made of yet smaller particles — the quarks — whose intrinsic size is smaller than we can yet measure. All that we can say for sure is that a single quark is no bigger than 1/10,000 the diameter of a proton. The same is true for the "planetary" electron relative to the proton "sun:" 1/10,000 rather than the "mere" 1/100 of the real solar system. So the world within the atom is incredibly empty.

So how does matter appear to be so solid? The atom may be empty as far as its particle content is concerned, but it is filled also with powerful forces. The electric and magnetic fields within an atom are far stronger than any we can make with even the most powerful magnets. It is these electromagnetic fields of force that make the atom so impenetrable and which are preventing you from sinking to the centre of the Earth as you read this.

In comparing an atom to the solar system, the matter of hundreds versus ten thousands is a minor issue compared to the real problem, which was this: such an atom should not be able to exist! According to the laws of physics, orbiting electrical charges such as electrons should radiate energy and spiral down into the nucleus within a fraction of a second. The fact that atoms have survived for billions of years shows that this cannot be the case, so what was going on?

It was Niels Bohr, the Danish physicist who was visiting Rutherford's group at Manchester University, who realised that previous experience with large-scale physical systems might be totally inadequate for dealing with microscopic systems, and came up with his atomic model. Indeed there were precedents for such caution. In 1900, Max Planck had shown that light is emitted in distinct microscopic 'packets' or 'quanta' of energy known as photons; and in 1905, Einstein showed that light remains in these packets as it travels across space.

The discovery that radiant energy is quantised led Bohr to propose that the energies of electrons in atoms are also quantised: they can have only certain prescribed energies. Restricted to these particular energy states, electrons do not radiate energy continuously and so do not smoothly spiral inwards. Instead, they can only jump from one energy state to another and emit or absorb energy to keep the total amount of energy constant (over long time-scales, energy is conserved) (see Table 2.3).

Table 2.3 "The uncertainty principle"

It is not possible to measure both the position and momentum of a particle with arbitrary precision. To observe an electron, we may shine light on it and detect the scattered radiation (photons). In the act of scattering photons, the electron recoils and alters its momentum. In the macroscopic world, this is of no consequence as the momentum of a massive object is not measurably affected, but for atomic and subatomic particles, the inability to determine precisely both spatial position and momentum is a fundamental problem.

If the position of a particle is known to be within some distance r of a point, then its momentum must be indeterminate by at least an amount p, where

$$p \times r \sim \hbar$$

and h is a constant of nature known as Planck's constant.

Many phenomena, such as angular momentum, are controlled by $h/2\pi$, which is denoted by $\hbar$. Its magnitude is

$$\hbar = 1.05 \times 10^{-34} \text{ Js } = 6.6 \times 10^{-22} \text{ MeVs}$$

Table 4.1 describes the essential role that Planck's constant plays in determining the size of atoms.

A similar uncertainty applies to time and energy. This implies that energy conservation can be "violated" over very short time scales. I put "violated" in quotes because one cannot detect it; this is the nub of the inability to determine energy *precisely* at a given time. Particles can radiate energy (e.g., in the form of photons) in apparent violation of energy conservation, so long as that energy is reabsorbed by other particles within a short space of time. The more that the energy account is overdrawn, the sooner it must be repaid: the more you overdraw on your bank account, the sooner the bank is likely to notice, but pay it back before being found out and everyone is satisfied. This "virtual" violation of energy conservation plays an important role in the transmission of forces between particles (see Figure 4.1 and the pion p. 53) and in sensing the existence of particles whose masses are too large for them to be produced (e.g., the top quark and the Higgs boson in Chapters 10 and 12).

In his theory of energy quanta, Planck introduced a quantity known today as 'Planck's constant,' traditionally abbreviated to the symbol h (the combination $h/2\pi$ being denoted $\hbar$). Bohr proposed that the permissible energies of electron orbits in atoms were controlled by the same quantity h. He then applied his idea to the simplest atom, hydrogen, whose nucleus is encircled by just one electron. No energy is radiated when the electron stays in an orbit, but if it jumps from a high energy to a lower energy state, then energy is emitted. Assuming that this radiated energy was converted to light, Bohr calculated the corresponding wavelengths and found that they matched precisely the mysterious spectrum of hydrogen. Planck's quantum theory, applied successfully to radiation by Einstein, had now been applied to matter with equal success by Bohr (see Figure 2.2).

The spectrum that an atom produces changes when magnetic fields are present. Some single lines may split in two, or be shifted slightly relative to others as the magnetic forces disturb the atomic electrons. These subtieties were all explained by supposing that not just the energy but also the angular momentum of the orbiting electron are quantised: restricted to integer multiples of Planck's constant $\hbar$. Furthermore, the electron was discovered to have an intrinsic spin of its own, of magnitude $\hbar/2$. All of this followed naturally once Bohr's hypothesis of 'allowed orbits' was accepted,

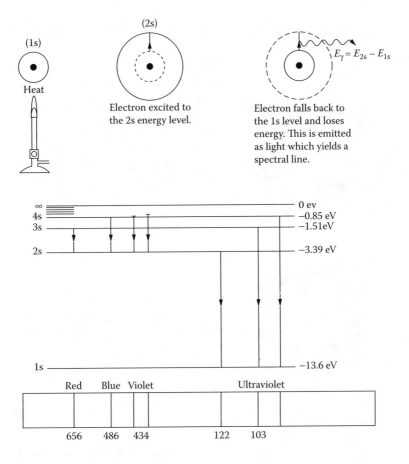

FIGURE 2.2 "The hydrogen spectrum." Energy levels in hydrogen for an electron in s-states, and some possible electron jumps that yield spectral light. The numbers denote the light's wavelength in nanometres: 1 nm is 10^{-9} m. Visible light spans the range 350 to 700 nm corresponding to energies of about 3.4 to 1.7 eV.

but in turn raised the question as to what ordained that electrons should choose these special states.

The essential clue lay with Planck's quantum theory that had been Bohr's inspiration in the first place. Planck and Einstein had shown that radiation with its well-known wave-like character could act like a staccato burst of particles called photons. Some 20 years later, in 1925, Louis de Broglie proposed the converse: particles of matter can exhibit wave-like characteristics. Planck's theory, as generalised by de Broglie, required that the wavelength of a low energy electron would be larger than that of a high energy one. Now imagine an electron circling a nucleus. The lowest energy Bohr orbit ('first orbit') contains exactly one wavelength; as we pass around the Bohr orbit we see the wave peak, then fall to a trough and back to a peak precisely where we started. The second orbit contains exactly two wavelengths, and so on. When

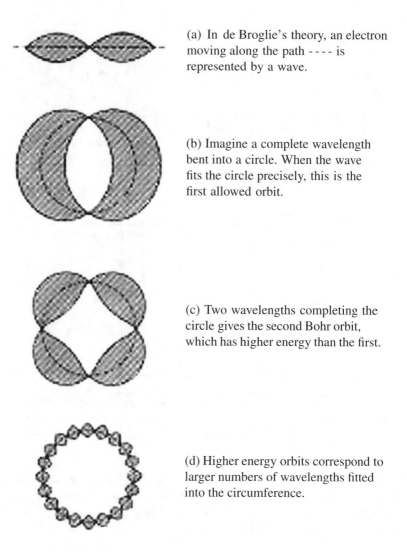

(a) In de Broglie's theory, an electron moving along the path - - - - is represented by a wave.

(b) Imagine a complete wavelength bent into a circle. When the wave fits the circle precisely, this is the first allowed orbit.

(c) Two wavelengths completing the circle gives the second Bohr orbit, which has higher energy than the first.

(d) Higher energy orbits correspond to larger numbers of wavelengths fitted into the circumference.

FIGURE 2.3 The de Broglie waves and allowed orbits in Bohr's atomic model.

the de Broglie electron wave exactly fits into a Bohr orbit, the wave reinforces itself by constructive interference — peak matches peak and adds — and so persists (see Figure 2.3). On the other hand, if the de Broglie wave does not fit into the orbit, the wave interferes destructively — peak meets trough and cancels out — and dies out rapidly. Thus discrete orbits and discrete energy states emerge directly from de Broglie's hypothesis on the wave nature of electrons.

This strange marriage of classical ideas with wave-like ingredients was both successful and disturbing. Great efforts were made to understand its workings more deeply, which culminated in the modern quantum theory developed by Schrodinger,

Heisenberg, Dirac, and others from 1928 onwards. The history of this is a fascinating story in its own right, but this is as far as we need to go in the present context. Modern quantum mechanics gives a more profound description of the atom than we have described but contains within it ideas that correspond to the classical images of 'solid' electrons orbiting and spinning. These conceptually helpful pictures are widely used even today, and I shall do so here.

3 The Nucleus

NEUTRONS AND PROTONS

The positively charged atomic nucleus is the source of intense electric forces, which are felt throughout the atom. On the one hand, these hold the negatively charged electrons in place, in the periphery of the atom; on the other, they repel positively charged invaders, such as alpha particles, violently ejecting them from whence they came. It had been this dramatic repulsion of the alpha particles that had first hinted at the existence of the massive, compact positive centre in the atom. The problem now was to see inside the nucleus, or even to get a direct look at it, as these same electric forces protected it from easy entry. So although the existence of a massive compact heart to the atom was established, its nature remained a mystery for nearly 20 years.

This slow progress contrasts with the profound understanding that had emerged about the outer reaches of atoms. By 1930, the structure of the atom was common knowledge. The rules governing the distribution of the electrons among the atomic energy levels had been successfully worked out and explained the regularities in chemical properties that Mendeleev had noticed 60 years earlier. The whole of theoretical chemistry was subsumed in the new atomic physics. This caused one of the most prominent theoretical physicists of the time to comment that the basic laws were now formulated and that the prime task was to work out their consequences.

That was not the first time, nor probably will it be the last, that experiment would show that a theoretical description of things then known is not the same as an explanation of all that there is to know. Nature has a habit of surprising us by being richer than we imagined. So it would be with the atomic nucleus. Far from being a mere passive source of the electric fields that ultimately guide the electrons in their choreography, controlling chemistry, biology, and even life, the nucleus turned out to have a labyrinthine structure of its own. Decoding that and investigating its implications would lead to a new science: nuclear physics, whose implications are still being worked out.

To learn more about the nucleus, Ernest Rutherford and James Chadwick bombarded many different elements with alpha particles in a series of experiments between 1919 and 1924. To begin with, they used nitrogen atoms as targets. The nitrogen gas filled a tube. They fired the alpha particles into one end of the tube; the nuclei of the nitrogen atoms scattered the alpha particles and at the far end the two scientists detected the results. Rutherford was amazed to find not just alpha particles, but also that the nuclei of hydrogen atoms emerged. This was remarkable because there had been no hydrogen there to start with: the alpha particles had somehow ejected hydrogen nuclei out of the nitrogen target. Similar things happened when they bombarded other elements, showing that the nuclei of one element could be changed into those of others. This suggested that nuclei are not featureless dense balls of positive charge but have some shared internal structure. The challenge was to find out what this is.

Hydrogen atoms are the simplest, consisting of just one electron and a nucleus which is nothing more than a positively charged particle called a proton. The proton's positive charge exactly counterbalances the electron's negative so that this atom has no net charge. However, in mass there is a gross asymmetry: the proton is 1836 times more massive than the electron and so provides the bulk of the hydrogen atom's mass. As the alpha particle bullets ejected protons from nitrogen and also from other elements, Rutherford proposed that it must be protons that carry the positive charge of *all* atomic nuclei.

The spectrum of x-rays and of visible light emitted by the atomic elements, and the underlying theory rooted in the new quantum mechanics, all fitted perfectly with the idea that as one moves up the periodic table of the elements, from one element to the next, so the amount of positive charge on the central nucleus increases by one unit at a time. In Rutherford's model, the more protons there are in a nucleus, the greater its positive charge; thus hydrogen has one proton, helium two, oxygen eight, and so on.

So far, so good. However, there was a problem. Whereas eight protons in an oxygen nucleus explained its positive charge, measurements of the relative atomic masses of the chemical elements implied that an oxygen atom is 16 times more massive than a hydrogen atom. Its eight protons provide only half of the mass: what contributes the remainder?

Whatever it was must have no electric charge and a mass similar to that of a proton. The simplest guess was that a proton and electron somehow grip one another tightly inside the nucleus, playing the role of an effectively neutral particle. However, this did not explain all the facts. Atomic nuclei can spin, and only at certain specific rates, according to quantum mechanics. Their spin affected the way that they behaved in magnetic fields and from this it was possible to measure the rate at which the nuclei of various elements spin. In the case of nitrogen, for example, the results could only be explained if its nucleus contained an even number of constituents in all. The proton-electron model would have required 14 protons to explain the mass and 7 electrons to give the net charge, which totals 21 particles, an odd number, and inconsistent with the even number required by the spin measurement. Rutherford then came up with the answer.

He guessed that instead of a proton and electron joining together, there might instead be a single genuine particle as heavy as a proton but with no electrical charge: he called it a 'neutron.' Neutrons contribute to the mass of the nucleus but not to its electric charge. This simple idea balances all the sums at once. Replacing the 7 proton and electron pairs by 7 neutrons gives the same charge and mass as before but now involves a total of 14 particles, an even number as required by the spin. The spin is satisfactorily described if the neutron spins at a rate identical to the proton. The same was true for the other elements. So the idea was economical and precise: all atomic nuclei are built from protons and neutrons. The challenge now was to prove it and the first step would be to demonstrate the neutron's existence.

Irene Joliot Curie, daughter of Marie Curie and Frederic Joliot, had evidence for the neutron but misinterpreted it. The neutron was discovered in 1932 by Chadwick. This is how it came about.

The Joliot-Curies had fired alpha particles at beryllium, the fourth lightest element, and discovered that electrically neutral radiation came out. Today we know that these were neutrons; however, mistakenly they thought them to be x-rays. When Rutherford heard of their results, he realised that they had probably inadvertently produced neutrons. He had discussed the idea of the neutron with Chadwick over many long hours when they were experimenting together and now it was to be Chadwick's great moment. He first repeated what the Joliot-Curies had done, and then added an essential ingredient to the experiment: he placed paraffin wax some distance away from the target and on the far side from the beam of alpha particles. The alphas hit the beryllium and ejected the mystery radiation; this radiation then hit the paraffin and, to Chadwick's delight, ejected protons from it. Rutherford compared this to H.G. Wells' invisible man: although you could not see him directly, his presence could be detected when he collided with the crowd. Thus it was when the invisible radiation collided with the paraffin wax.

X-rays have no mass and would have been unable to do this under the conditions of Chadwick's experiment. Their energy would easily remove lightweight *electrons* from the atoms in the paraffin but would not eject *protons*; the proton is so massive that it would merely shudder under the impact. Whatever was knocking out the proton must itself be heavy. Chadwick suggested that here was evidence for a new subatomic particle, similar in mass to the proton but with no electrical charge: the neutron that Rutherford had predicted. Apart from a one part in a thousand difference in mass and the presence of electrical charge, the proton and neutron are identical. As they are constituents of the nucleus, they are often referred to collectively as 'nucleons.'

Every nucleus of a given element contains the same number of protons but may have different numbers of neutrons. Hydrogen usually has one proton and no neutrons, but about 0.015% of hydrogen atoms consist of a proton and a neutron. This is known as the *deuteron*, the nuclear seed of deuterium, sometimes called 'heavy hydrogen.' A proton accompanied by two neutrons is known as the 'triton,' the seed of tritium.

The total number of neutrons and protons is placed as a subscript to the atomic symbol to distinguish the various isotopes. Thus, $^{92}U_{235}$ and $^{92}U_{238}$ are two isotopes of uranium: both have 92 protons but contain 143 and 146 neutrons, respectively, and hence 235 and 238 'nucleons.'

In 1932, the same year that Chadwick discovered the neutron, nuclei were split for the first time by artificial means. Whereas previously, alpha particles produced by the natural radioactive decays of radium had been used as probes, now John Cockroft and Ernest Walton used electric fields to accelerate protons to high speed, and then fired these beams of high-energy particles at lithium nuclei. This had two advantages over what had been done before. First, protons with their single positive charge feel less electrical resistance than do the doubly charged alpha particles when approaching a nucleus. Second, and most important, the high-speed particles penetrate deeper before being slowed. Cockroft and Walton had thus made the first nuclear particle accelerator, and created a practical tool for investigating within the atomic nucleus. This was the prototype of the modern particle accelerators that have been used for probing the internal structure of the neutrons and protons themselves.[1] (See Table 3.1.)

---

[1] For more about accelerators, see *Particle Physics; A Very Short Introduction* by Frank Close.

Table 3.1 Early particle accelerators

Electric fields will accelerate electrically charged particles such as electrons or protons. Röntgen's 1895 x-ray tube accelerated electrons by a potential of about a thousand volts. When these electrons were decelerated, for example, by hitting a screen, electromagnetic radiation (such as x-rays) was emitted. In modern times this is a principle behind the television.

Accelerate electrons along the full extent of a longer tube and they can reach higher energies. Present technology can increase an electron's energy by more than 30 MeV per metre of tube. The Stanford Linear Accelerator (SLAC) in California is 3 km long and the electrons attained energies of up to 50 GeV. At such energies, electrons can smash into nuclei and even probe the inner structure of protons. This increase in the size of electron accelerators over the years has been matched by proton accelerators.

Rutherford's early work on the atomic nucleus in 1911 used beams of naturally occurring alpha particles that had been emitted by radioactive nuclei. A controllable source of high-energy particles was made by Cockroft and Walton in 1932. This accelerated protons by a potential of 500,000 volts, whereupon they were smashed into nuclei, disintegrating them.

To accelerate protons to higher energies required larger distances over which the accelerating force could be applied. Ernest Lawrence was the first to accelerate protons around a circular orbit. A magnetic field bent their path around a semicircle; an electric field gave them a kick; then they were bent round a second semicircle, and so on. The protons circled round and round and reached an energy of 1 MeV. This device was called a "cyclotron." It has long been superceded, but the idea of using a combination of magnetic fields, which steer the particles, and electric fields, which speed them, underpins modern accelerators. Circular machines many kilometres in circumference accelerate electrons to energies of hundreds of GeV or protons to greater than 1000 GeV, known as 1 TeV.

RADIOACTIVITY AND NUCLEAR FISSION

Protons and neutrons are the common ingredients of *all* nuclei and so one variety of nucleus can transmute into another by absorbing or emitting these particles.

A common example involves the source of the alpha particles that had been used to such great effect by Rutherford and colleagues in decades of experiments. An alpha particle consists of two protons and two neutrons tightly clumped together. So in isolation this combination forms the nucleus of helium. This little cluster is so compact that it almost retains its own identity when buried in a large number of protons and neutrons such as the nucleus of a heavy element. Sometimes the heavy nucleus gains stability by spontaneously ejecting the quartet. The net numbers of protons and neutrons are separately conserved throughout: one cluster has broken into two. This spontaneous decay of nuclei is an example of *radioactivity* and is the explanation of Becquerel's 1896 discovery of the phenomenon.

In the conventional description, the alpha particle is $^2\text{He}_4$. The emission of an alpha particle, for example, when uranium nuclei break down into thorium, is summarised by:

$$^{92}\text{U}_{238} \rightarrow \,^{90}\text{Th}_{234} + \,^2\text{He}_4$$

Not all nuclei are radioactive: for all practical purposes, most are stable. The most stable nuclei tend to be those where the number of neutrons does not greatly exceed the number of protons. One of the exciting projects in the early 1930s was to bombard naturally occurring nuclei with neutrons, in the hope that some neutrons would attach themselves in the target nuclei and form new *isotopes*. Neutrons having no electrical charge are not repelled as they approach a nucleus and so can gain more easy access than protons. This made them especially useful in nuclear investigations. To prevent the neutrons from hitting the nucleus hard and shattering it, Enrico Fermi first slowed them by passing them through paraffin. With this technique he successfully modified the nuclei of various atoms. He attached neutrons to fluorine, producing a new artificial isotope of that element, and did likewise with a total of 42 different nuclear targets until, in 1934, he came to the heaviest known element: uranium. Fermi was one of the greatest physicists of all time, but at this point he made a mistake with potentially historic implications.

Irradiating the uranium gave him some puzzling results which suggested that the neutron had not simply become attached to the uranium nucleus. Fermi thought that the anomalies might be evidence that he had produced the first *transuranic* element, one place above uranium in Mendeleev's table, unknown on Earth but capable of existence in principle. To have done so would have been a great prize and he was so taken up with this possibility that he missed the real explanation.

Otto Hahn and Lisa Meitner in Germany repeated the experiment. Meitner was Jewish and in consequence fled from Germany to Sweden, while Hahn continued with Fritz Strassmann and analysed the chemistry of what had been produced. This showed that far from new elements having been made, the products contained familiar stuff such as barium, an element with a nucleus of 56 protons. This was a great surprise: could slow-moving neutrons so disturb uranium that it split into two almost equal halves?

Hahn told Meitner, now in Sweden, of this. She discussed it with Otto Frisch and together they came up with the explanation. The nucleus is like a liquid drop. Liquid drops are held together by surface tension, the nucleus by the strong force. The electrical charge of the protons in a nucleus works against the strong force, and the heavier the element, the bigger the repulsion. Beyond uranium the two forces work against each other and cancel such that no stable elements occur. Uranium itself is so delicately balanced that slow neutron bombardment makes a uranium nucleus wobble like a liquid drop and break up:

$$n + \,^{92}\text{U}_{235} \rightarrow \,^{56}\text{Ba}_{144} + \,^{36}\text{Kr}_{89} + 3n$$

In the fragments there are further slow neutrons that can trigger the break-up of other uranium nuclei, and enormous energies can be released in the ensuing *chain reaction* (Figure 3.1). The essential ingredients of the so-called atom bomb were unwittingly at hand in Italy and Germany up to 5 years before World War II.

1. A slow neutron is absorbed by a uranium-235 nucleus.
2. The uranium is now unstable and wobbles like a water drop.
3. The nucleus becomes so deformed that it splits in two.
4. The products are stable nuclei of barium and krypton and two or
   three neutrons. Energy is also released.
5. One of these neutrons might hit another fissionable nucleus of uranium
   so that a chain reaction develops. Energy is released explosively if more
   than one neutron induces another fission.

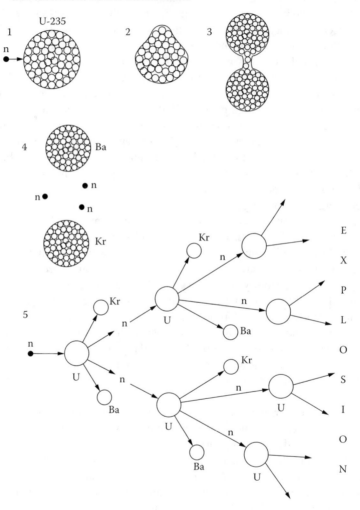

FIGURE 3.1 Slow neutrons disintegrating uranium.

The splitting of nuclei in this way is called *nuclear fission*. If Fermi had seen the answer in 1934 or if Meitner had not fled from Germany, then the atom bomb might have been developed by Hitler's scientists. As it was, Fermi also fled, and played a major role in the Allies Manhattan project.

BETA RADIOACTIVITY AND NEUTRINOS

Becquerel discovered three types of radiation in 1896 (Figure 3.2): alpha particles are helium nuclei and gamma rays are extremely energetic photons emitted from the nucleus when the nucleons rearrange themselves and lose energy (see Figure 3.3). Beta-radiation consists of electrons: where did they come from?

Heating or irradiating atoms can supply enough energy to eject one or more of their outer electrons. However, the production of Becquerel's electrons was quite different: they emerged without energy being first supplied to the atom. Their source is not the electrons in the periphery of the atom; instead they are emitted by processes occuring within the nucleus itself (Figure 3.4).

Isotopes containing a high percentage of neutrons tend to be unstable. Not only can they spontaneously emit alpha particles, but also a neutron can spontaneously convert into a proton and emit an electron in the process. Electrical charge is always conserved in nature and this transmutation is a good example:

$$n^0 \rightarrow p^+ + e^-$$

The superscripts denote the electrical charges of n (neutron), p (proton), and e (electron). This neutron decay is known as β (beta) radioactivity and is the source of many nuclear transmutations.

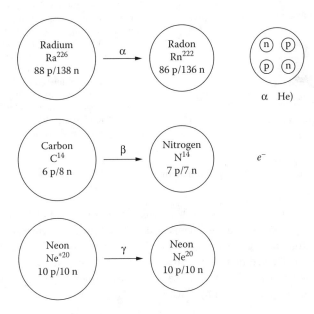

FIGURE 3.2 Three varieties of nuclear radioactive decay. Alpha (α) and beta (β) decays involve a change in the neutron and proton content of the nucleus, and change it to a different species. Following either of these decays, the protons and neutrons rearrange themselves and in the process emit energy in the form of gamma (γ) rays. No disintegration occurs in gamma emission: one or more nucleons is temporarily in a state of high energy ("excited nucleus"), which it loses by emitting a gamma ray. In the figure, the excited nucleus that existed before gamma emission is labelled with an asterisk.

Atoms

Spectra

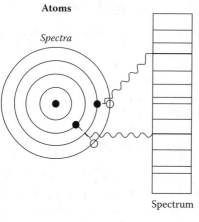

In Bohr's model of the atom, light is emitted when an atomic electron jumps from one orbit to another. The energy that the light carries off is equal to the difference of the energies of the electron before and after:

$$E = E_{\text{initial orbit}} - E_{\text{final orbit}}$$

Different energies of light have different colours.

A spectrum gives information on the nature of the energy levels available for the electrons. The colours reveal the differences in energies and from these the actual set of energy levels can be deduced.

Spectrum

X- rays

Heavy atoms have lots of electrons. If an electron drops to the lowest energy level all the way from the highest ones, then a lot of energy is emitted as electromagnetic radiation. These energetic bursts of radiation are x-rays.

The nucleus

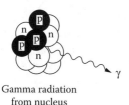

Gamma radiation
from nucleus

The nucleus at the atom's heart has a complicated structure of its own. It contains *neutrons* and *protons*. When protons change from one type of motion to another in the nucleus, very high-energy electromagnetic radiation is sometimes emitted : this is called *gamma rays* (denoted γ). This is analogous to the way that visible light is emitted in atomic rearrangement of electrons.

FIGURE 3.3 How photons emerge from atoms.

One puzzling feature was that in beta-decay, the proton and electron seemed to have less energy than they ought. Energy is conserved over long time scales but can be converted from one form, such as potential, kinetic, chemical, or heat energy, into another. Einstein showed that another form of energy conversion can occur: energy can be converted into mass and vice versa. The amount of energy (E) produced when a mass (m) is destroyed is given by Einstein's equation:

$$E = mc^2$$

where c is the velocity of light. An isolated neutron has slightly more mass than a proton; in energy this corresponds to some 1.5 MeV in 940 MeV. According to Einstein's relation $E = mc^2$, the energy released in the decay is:

$$E = m(neutron)c^2 - [m(proton)c^2 + m(electron)c^2] = 0.8 \text{ MeV}$$

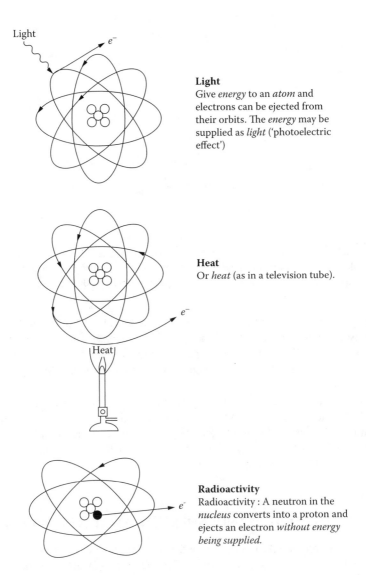

Light

Light
Give *energy* to an *atom* and
electrons can be ejected from
their orbits. The *energy* may be
supplied as *light* ('photoelectric
effect')

Heat
Or *heat* (as in a television tube).

Radioactivity
Radioactivity : A neutron in the
nucleus converts into a proton and
ejects an electron *without energy
being supplied.*

FIGURE 3.4 How electrons emerge from atoms.

(see "Energy accounts: from macro to micro" at end of Chapter 1) which should
manifest as kinetic energy of the proton and electron. Energy seemed to have unac-
countably disappeared, which would be contrary to centuries of evidence that energy
is conserved over long time scales. Not only did energy seem to be unaccounted for,
but the momentum of the emerging proton and electron did not add up correctly either.
If a stationary neutron breaks down into two particles, then they should move off in
opposite directions along a straight line. However, the proton and electron moved off
at an angle, as if some unseen third particle was also being produced in the decay.
This invisible particle was responsible also for the apparent energy imbalance as it
takes away some of the energy itself.

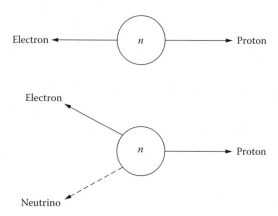

FIGURE 3.5 The neutrino. If a neutron decayed to two particles, they would move off back-to-back along a straight line as in the first diagram. In practice they were seen to move off at an angle to one another due to a third (invisible) particle being produced. This is the *neutrino*.

The particle carries no electrical charge and a mass that is trifling at most. Wolfgang Pauli postulated its existence in 1931 to explain the otherwise anomalous energy loss in the neutron decay, but it was not detected directly until 1956. It was named the 'neutrino' (to distinguish it from the neutron) and it is conventionally denoted by the symbol ν (see Figure 3.5).

Neutrinos are ejected in nuclear transmutation but have no existence within a nucleus. As such, they are like the electron produced by β-decay. We will meet the neutrino in Chapter 12. In the present chapter, we are concerned with the neutrons and protons that make nuclei. We need to understand why some nuclei are stable and others are not. This is what we do next.

NUCLEAR STABILITY AND WHERE THE ELEMENTS CAME FROM

Atomic nuclei are compact, still somewhat mysterious entities consisting of up to a couple hundred protons and neutrons. The neutron is neutral, in that it has no overall electric charge, while the proton is positively charged. Apart from this manifest difference, the two are otherwise almost identical. They have essentially the same shape, size, and mass (or almost; the neutron is a trifling one part in a thousand heavier than a proton). When neutrons and/or protons touch one another, they grip together tightly. This force, which ultimately builds atomic nuclei, is known as the *strong force*. The strength of its grip between pairs of neutrons is the same as between protons, or indeed between a proton and a neutron. Two protons will feel a slight electrical repulsion ("like charges repel") which neutrons, or a proton and neutron conjoined, will not feel. So, unlike neutrons, protons are pushed slightly apart from one another due to the electrostatic force. As far as the strong force is concerned, for our present purposes we can think of protons and neutrons as miniature versions of billiard balls that are distinguished solely by being white (the neutron, say) or red (the proton) and imagine what happens if at random we pick balls from a bag containing essentially an infinite number of them.

This scenario is a metaphor for what we believe occurred 5 billion years ago. (Part of this story will substantiate this belief.) The bag is akin to the heart of a long-dead star.

Stars such as our Sun, large glowing spheres held together by gravity, consist mainly of hydrogen and helium with smaller amounts of other elements. They are hot and dense enough in their central core for nuclear burning — fusion — of light elements to take place, converting them into heavier ones. This liberates energy, with the result that the star's life involves a battle between this outward pressure and the inward pull of gravity. Initially, stars fuse hydrogen into helium. The helium settles in the central core of the star until the helium nuclei also start fusing together, building up the nuclear seeds of heavier elements. This cookery produces yet more heat, causing the star to swell into what is known as a *red giant*, which is what our own sun will become in another 5 billion years. After that, with all its helium gone and unable to resist gravity's force, it will shrink under its own weight and form a *white dwarf*.

Stars with about ten times more mass than the Sun can carry on burning, producing carbon, oxygen, and elements in the periodic table as heavy as iron. In heavier stars, once the nuclear fuel has been used up, there is no heat pressure to resist gravity and the dense iron core collapses catastrophically; the outer layers fall in on the core and bounce off, sending gigantic shock waves outwards. As these pass through the outer regions of the infalling star, further nuclear reactions take place, which create elements heavier than iron. Momentarily they also make huge unstable balls of neutrons and protons gripped into highly unstable nuclei heavier than even uranium.

The exploding star shines brightly — this is what we call a *supernova*. Left behind is a compact neutron star or a black hole; emitted into space are the nuclear seeds of the periodic table, a host of nuclear lumps, balls of protons and neutrons in assorted combinations, along with electrons.

In the coolness of space, the positively charged nuclei attract the electrons to form atoms and collectively mutually attract one another by gravity to form huge balls, such as planet Earth. Today we find the elements of the periodic table ranging from hydrogen, whose nuclear seed contains but a single proton, to uranium with 92 protons and anything from 140 to 150 neutrons. There are two immediately noticeable things here. First, apart from hydrogen, the naturally occuring nuclei tend to contain more neutrons than protons (up to about 50% more, as in uranium). Second, that although all numbers of protons are found up to 92, some are more common than others (such as iron with 28 protons, whereas astatine with 85 is almost unheard of), and beyond 92, the combinations such as plutonium, berkelium, and einsteinium have been made only artificially, as in nuclear laboratories or explosions. If neutrons and protons are so similar, why are some combinations as common as the oxygen in the air we breathe and others are rarer than gold?

First, let us be sure we have our metaphor agreed. An atomic nucleus in this model is like a collection of billiard balls. Any **A**tomic nucleus is then made from a total of **A** such balls. This number, selected at random from the stellar bag, will consist of white balls (**N**eutrons, number **N**) and red balls (**P**rotons, number **P**); obviously $A = N + P$. Any balls that have touched one another will stick together.

Imagine what would happen if you dipped into the bag and pulled out bundles at random. We would expect that there will be many clumps where the number A is small; fewer where A is large, but we would hardly expect to find a sudden end at 92, or any other number; we would expect a gradual falling off in the chance of finding ever larger numbers. We would also expect to find white balls as often as red, and hence that clumps will, on average, contain similar numbers of red and white balls; in other words, that $N \sim P$. This is indeed similar to what we find in the real world. The nuclear "isotopes" (that is, the way that for a given number P, how many N and hence A are found) do not occur when N and P differ by large amounts. However, there is a clear tendency for N to be somewhat larger than P, which makes sense, as we hinted already: the red balls (protons) are marginally less likely to stick to one another due to their mutual electrical repulsion.

At this point you might wonder why there are not any nuclei containing neutrons alone; if protons add electrical instability, why do neutrons not cluster on their own? The answer lies in their one part in a thousand extra mass. Adding neutrons costs a little in mass, and Einstein's $E = mc^2$ implies that mass equates to energy. So it costs energy to add neutrons (due to their mass) and it costs energy to add protons (due to their electrical repulsion). There is thus a competition, and the balance favours the neutrons slightly. This qualitatively explains why the elements have N larger than P, but not by too much.[2]

This extra energy locked into the neutron's mass also leads to instability. As nature seeks the state of lowest energy, like water running downhill to sea level, so a neutron left to itself will eventually experience beta-decay: $n \rightarrow pe^- \bar{v}$, which converts it to the marginally lighter proton, the excess energy being transformed into the electron and neutrino. So while a neutron in a nucleus can be stabilised (see the next section), if you gather too many together, they will undergo beta decay, increasing the number of protons at the expense of neutrons. Conversely, try putting too many protons together and their electrostatic repulsion destabilises them; in this case it is possible to lower the net energy by "inverse beta decay" where one of the protons converts to a neutron: $p(\text{in nucleus}) \rightarrow n(\text{in nucleus}) + e^+ + v$. The net effect is that when collections of N and P get too big a mismatch, beta-decay or inverse beta-decay moves the whole back towards the "valley of stability" where the number of neutrons N tends to exceed the number of protons P.

The tendency for neutron excess gets more marked as one moves to ever larger clusters. Thus for a stable isotope of barium, say, where $A = 144$, $P = 56$, and $N = 88$, the excess is some 45%; whereas by uranium with $A = 235$, $P = 92$, and $N = 143$, the extra 36 protons have been joined by a further 55 neutrons, leading to an excess of over 50%. This overpopulation of neutrons is part of the reason for the chain reaction that results when neutrons cause U_{235} to fission. If a stray neutron impacts on U_{235} and splits it in two, this leads to the nuclei of lighter elements such as barium. The number of protons adds up to 92, of course, but relatively fewer neutrons are needed in the stable forms of these lighter nuclei than in heavy uranium. So not all of uranium's neutrons are needed to make the barium and krypton. These "spare"

---

[2] There is one exception. When $N > 10^{43}$, neutrons can hold one another by their mutual gravity. Such a "neutron nucleus" extends over s few kilometres and is what we call a *neutron star*.

neutrons are thus free to roam and perhaps hit other uranium nuclei, splitting them ("fission") and in turn releasing yet further spare neutrons. Not only does this release neutrons, but it also releases energy. Unless the neutron emission and the fission are somehow stopped (by the presence of other elements such as carbon that acts as a sort of blanket), the process can grow explosively.

Beyond uranium, all combinations are unstable. They can reduce their net energy, thereby increasing stability, by means of beta decay or alpha emission. When an alpha particle (consisting of two protons and two neutrons) is ejected from a large nucleus, this tends to produce a more stable remnant. Nuclei beyond uranium, and many even smaller than it, eject these alpha-particle combinations and produce more stable end-products. For nuclei that are bigger than uranium, such spontaneous decays occur very rapidly, often within fractions of a second. Uranium itself is not stable, though its half-life (that is, the time that half of a sample will have spontaneously decayed) · is some 5 billion years, as old as the Earth. This means that about half of the uranium that existed at the birth of our planet has by now decayed. It is by measuring the relative abundances of the various isotopes of uranium and other heavy elements that we have been able to deduce the age of the Earth. Billions of years into the future, nearly all of the uranium will have decayed away, leaving lead as the most bulky naturally occurring nucleus.

We are now in a position to describe how the elements on Earth were formed. (See also Table 3.2.)

A supernova explosion 5 billion years ago produced nuclear clumps of all sizes. Many small lumps had formed, seeding the elements that are lighter than lead. Once in the relative coolness of this newborn Earth, the superheavies decayed, cascading rapidly into smaller lumps, the largest of which were the nuclei of uranium atoms. The uranium also decayed, albeit slowly (half-life of billions of years) and is still doing so.

Once through this uranium barrier, the decay products of uranium themselves decay rapidly through a variety of chains, each of which eventually reaches lead. Lead is the heaviest permanently stable element. Among the uranium to lead chain are some elements whose half-lives are so brief, and the chance of them being formed in the first place is so small that their net abundance on Earth is trifling: thus, for example, astatine and francium, which at any moment total less than a gramme in the inventory of the entire planet.

The reasons why some elements such as iron are common, while others such as gold are not, can involve details of quantum mechanics which go beyond this discourse. The result though is that some combinations of protons or neutrons are very stable. The numbers involved are known as *magic numbers.*

The magic numbers are predicted by quantum mechanics and are 2, 8, 20, 28, 50, 82, 126. So, for example, the alpha particle is "doubly magic" in that it contains two (a magic number) neutrons and two protons. Some examples of other combinations of magic numbers include 8 and 8 ($^8O_{16}$, oxygen); 28 and 28 ($^{28}Fe_{56}$, iron); and $^{82}Pb_{208}$ (lead; $208 = 82 + 126$!). These are each highly stable and very common in nature.

Quantum mechanics also gives an "affinity for pairs," namely, that a proton and a neutron have an extra strong affinity when numerically balanced. Nuclei that have an unbalanced number will tend to decay radioactively until the numbers pair off. So it

Table 3.2 Age of the earthly elements and the stars

The natural radioactivity of the elements in rocks is like a clock that started when the Earth was born. If we know how to read it, we can tell how old the Earth is.

When rocks solidified in the cooling new born Earth, their chemical elements often became separated into different crystalline grains within the rocks. Two radioactive elements, rubidium and strontium, are quite common and are usually found in different grains of a rock. Strontium, (Sr) has two particularly common isotopes: Sr_{87}, which is radioactive, and Sr_{86}, which is not. The chemical processes that separate elements into different grains of the rock do not separate isotopes: the ratio of Sr_{87} and Sr_{86} is the same in all grains.

Now let us imagine a grain of the rock that, when formed aeons ago, had no strontium and only rubidium, Rb_{87}. This is radioactive and decays into Sr_{87}, so more and more Sr_{87} will form as time goes by. Measurements in the laboratory show that this happens quite slowly, in fact we would have to wait for 47 billion years for half of the Rb_{87} to have decayed, but instruments are sensitive enough to measure this in even short timespans. So if we measure the ratio of Sr_{87} to Rb_{87} in the rock grain, we can determine how long the Rb_{87} has been decaying away, and hence determine the age of the rock, and ultimately the amount of time since the molten Earth cooled.

Unfortunately it is not so simple. It would be fine if there had indeed been no strontium there to begin with; if on the contrary there had been some, it will ruin our interpretation. What we need is some way of also measuring how much strontium was also there at the start. That we can do by measuring **both** the ratio of Sr_{87} to Sr_{86} and also that of Rb_{87} to Sr_{86}. From these we can work out how long the radioactive clock has been ticking. This sort of test can be done with other radioactive elements to check that the answers come out the same. They show that the oldest rocks are about 3.8 billion years old. Similar measurements have been made for meteorites, which turn out to be as old as 4.6 billion years. This is essentially the age of the solar system.

We can even measure the ratios of various isotopes in stars by the "autograph" of gamma rays that they emit. Different isotopes emit a spectrum of gamma rays of characteristic frequencies, or energies, such that they shine like beacons across the vastness of space and are detected here on Earth. Once we know from this what the relative abundances of the isotopes of various elements are, we can perform the same sort of calculations as for the Earthly rock samples and deduce how long the nuclear clock has been ticking in the star. This method shows that the age of the oldest stars is between 9 and 15 billion years. This is quite independent of estimates of the time since the Big Bang based on either the Hubble expansion rate or "direct" measurements from the WMAP satellite (Chapter 14). This illustrates just one of the many ways that the phenomenon of radioactivity provides a remarkable tool to reveal the most profound information about our origins.

is ultimately quantum mechanics that underpins the abundance of the elements in the natural world. We are stardust courtesy of radioactive decays leading to the stable end-products via the almost stable element of uranium. It is the differing half-lives of these unstable isotopes that over the epochs cause their relative abundances in materials to change. In turn, this enables us to determine the ages of rocks, the geological time-spans, and the age of the planet itself. This is all a result of neutrons and protons clinging together, but it is quantum mechanics that causes not all combinations to be favoured equally: some, such as $^2He_4$, $^8O_{16}$ and $^{28}Fe_{56}$, "are more equal than others."

A question that I am often asked is "how can neutrons be stable inside nuclei but unstable on their own?" If you want to know, read the next section; if not, you can skip it without losing the storyline.

NEUTRONS: WHEN ARE THEY STABLE AND WHEN NOT?

An isolated neutron at rest has an energy $E = m_nc^2$, where m_n is the neutron mass. An isolated proton likewise has $E = m_pc^2$, which is slightly less than that of the neutron. So when a neutron undergoes beta-decay $n \rightarrow pe^-\bar{v}$, it loses energy and ends up in a lower energy state — the proton. The proton is the lightest such nuclear particle and cannot lower its energy by turning into anything else. As a result, an isolated proton is stable. This enables the proton, the nuclear seed of the hydrogen atom, to exist in stars and in substances on Earth long enough that interesting things can happen, such as the fusion processes in the heart of the Sun that build up the heavier elements leading to billions of years of evolution, and us.

Now imagine that neutron in a nucleus, which contains also one or more protons. The simplest example is the deuteron, a form of the hydrogen nucleus consisting of a single neutron and a single proton. This is stable. Why is the neutron stabilised here?

To answer this, think what would be the end result were the neutron to decay. At the start we had a neutron and proton touching; at the end there would be two protons. So although the neutron would have lost energy in its conversion to a proton, the two protons that ensue would have a mutual electrical repulsion, which adds to the energy accounts. The amount of this electrostatic increase exceeds that lost in the $n \rightarrow p$ conversion. Overall, the net energy would increase in going from the $(np) \rightarrow (pp)$ pair and so the initial (np) is stable and survives.

This generalises to nuclei where there are similar numbers of neutrons and protons: beta-decay is prevented due to the extra electrical effects that would occur in the final "proton-rich" environment. However, try putting too many neutrons in one ball and eventually the downhill energy advantage of them shedding a little bit of mass on turning into protons is more than the uphill electrical disadvantage of the protons' electrical repulsion. So there is a limit to the neutron excess in nature: too many neutrons and we find beta-decay taking place.

The energy accounts that are at work in the neutron stability example also explain the phenomenon of "inverse beta-decay," or "positron emission." There are some nuclei where a proton can convert into a neutron, emitting a positive version of the electron (known as a positron and the simplest example of an antiparticle, or more generally antimatter; see Table 4.2). This basic process is represented $p \rightarrow ne^+v$.

This does not happen for a free proton as it would have to "go uphill," in that the energy $m_n c^2$ of the final neutron would exceed that of the initial proton due to its extra mass. However, in a nucleus where there are several protons, this inverse beta decay can *reduce* the electrostatic energy as there will be one less proton at the end than at the start. If this reduction in energy exceeds the price of replacing an $m_p c^2$ by the larger $m_n c^2$, then inverse beta decay, $A(P, N) \rightarrow A(P - 1, N + 1)e^+\nu$ will occur.

There are several examples of natural "positron emitters" in nature. They have great use in medicine. The PET scanner (positron emission tomography) exploits this phenomenon.

4 The Forces of Nature

GRAVITY AND ELECTROMAGNETISM

By the end of 1932 there was a general belief that the basic building blocks of matter had been isolated: electrons and neutrinos, protons and neutrons. The outstanding theoretical task was to formulate the laws governing their behaviour and thus explain how matter and the world about us are made.

If the basic particles are likened to the letters of the alphabet, then there are also analogues of the grammar: the rules that glue the letters into words, sentences, and literature. For the universe, this glue is what we call the *fundamental forces* (Figure 4.1 and Table 4.1). There are four of them, of which *gravity*, the force that rules for bulk matter, is the most familiar. Matter is held together by electric and magnetic forces. These are profoundly related by what is known as *electromagnetic force*; it is this force that holds electrons in atoms and links atoms to one another to make molecules and larger structures, such as you, me, and the Earth beneath us. So it is gravity that attracts things to the ground and the electromagnetic force that stops them falling all the way to the centre of the Earth. Within and around the nucleus we find the other two forces: one is stronger than the electromagnetic force, the other being weaker. As a result, they have become known as the *strong* and *weak forces* though, as we shall see, this can be something of a misnomer. The strong force glues the protons and neutrons in the nucleus; the weak force changes one variety of particle into another, as in beta-radioactivity. This quartet of forces controls our lives.

Electromagnetic and gravitational forces and the rules that they obey had been known for a long time. Gravity mutually attracts all matter whereas electromagnetic force can attract or repel. The familiar adage 'unlike charges attract, like charges repel' is a first step in understanding the structure of the atom; the electrical charge of an electron is negative, that of the nucleus is positive, so their mutual attraction holds the electron in the latter's electrical grip.

The electromagnetic force is intrinsically much more powerful than gravity, so why was gravity identified first? In bulk matter, positive and negative charges tend to cancel, cutting off electromagnetism's sphere of influence and leaving the all-attractive force of gravity as dominant. Gravity is particularly noticeable over astronomical distances. The power of the electromagnetic force is felt in those situations where the effects of the positive and negative charges do not precisely cancel. The most familiar example is where the orbiting and spinning motions of the charges in atoms give rise to readily observable magnetic effects. A small magnet can attract to itself a lump of metal, overcoming the downward pull that the metal is experiencing from the gravitational attraction of the whole earth. The swirling electric charges in the Earth's core create magnetic effects such that the planet is a huge magnet. A small compass needle will swing in line with the Earth's magnetic field, pointing to the north and south magnetic poles. On a small scale, such as between individual atoms, the effects of gravity are negligible and it plays no observable role in describing known atomic or nuclear phenomena.

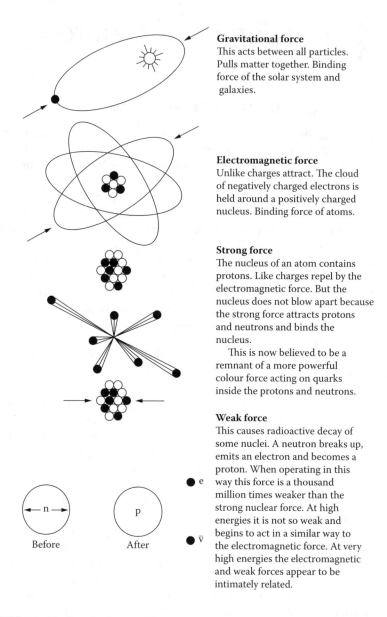

Gravitational force
This acts between all particles. Pulls matter together. Binding force of the solar system and galaxies.

Electromagnetic force
Unlike charges attract. The cloud of negatively charged electrons is held around a positively charged nucleus. Binding force of atoms.

Strong force
The nucleus of an atom contains protons. Like charges repel by the electromagnetic force. But the nucleus does not blow apart because the strong force attracts protons and neutrons and binds the nucleus.

This is now believed to be a remnant of a more powerful colour force acting on quarks inside the protons and neutrons.

Weak force
This causes radioactive decay of some nuclei. A neutron breaks up, emits an electron and becomes a proton. When operating in this way this force is a thousand million times weaker than the strong nuclear force. At high energies it is not so weak and begins to act in a similar way to the electromagnetic force. At very high energies the electromagnetic and weak forces appear to be intimately related.

FIGURE 4.1 The four fundamental forces.

In the 19th century, various electric and magnetic effects had been recognised for some time, but they were still not well understood. There were a variety of ideas for one situation or another. Coulomb's law described the behaviour of electrical charges in the presence of other electrical charges; the law of Biot and Savart described the force between two wires carrying electrical currents. A collection of seemingly independent laws governed various other electrical and magnetic phenomena.

Table 4.1 Nature's Natural Scales

The gravitational force between two objects whose masses are m and M and a distance r apart is:

$$F = GMm/r^2$$

If the force F is measured in newtons; m and M in kilograms, and r in metres, then

$$G = 6.67 \times 10^{-11} Nm^2 kg^{-2}$$

G is a fundamental quantity, the gravitational constant, that has dimensions. Similarly the velocity of light ($c = 3 \times 10^8$ m s^{-1}) and Planck's constant h have dimensions, as does the magnitude of electrical charge ($e = 1.6 \times 10^{-19}$ coulomb). The numerical values of these quantities depend upon the units employed. (It is common to write $\hbar = h/2\pi$.)

From these fundamental constants of nature one can form numerical quantities that are dimensionless or have dimensions of length. Do these have a deep significance?

Examples are the ratios of masses of the elementary particles, for which there is yet no explanation. Another interesting dimensionless quantity is

$$e^2/4\pi\hbar c \sim 1/137$$

This involves electric charge, e, the $\hbar$ of quantum theory, and the velocity of light c, which is important in relativity theory. Thus one might expect that this quantity will be important in a relativistic (c) quantum theory ($\hbar$) of electric charge (e).

Such a theory exists, known as "Quantum Electrodynamics." The above quantity is known as "α" and is the measure of the strength by which electrons couple to electromagnetic radiation.

If M and m are proton and electron masses, then the ratio of gravitational to electromagnetic forces in hydrogen atoms is

$$GMm/e^2 \sim 10^{-40}$$

which quantifies our statement that gravity is exceedingly feeble in atomic and particle physics. The gravitational constant, $\hbar$, and c form a quantity with the dimension of a length

$$\sqrt{(G\hbar/c^3)} \sim 10^{-35} \text{ m}$$

By analogy with quantum electrodynamics this suggests that the relativistic (c) quantum theory ($\hbar$) of gravity (G) becomes important at distances of order 10^{-35} m, or in energy equivalent, using the uncertainty principle, at 10^{19} GeV. This is far beyond the reach of present technology and gravity can indeed be ignored in present high-energy physics experiments.

Other important quantities with dimensions of length include

$$\hbar/m_\pi c \sim 10^{-15} \text{ m}$$

and

$$\hbar/m_e c\alpha \sim 10^{-10} \text{ m}$$

The former suggests that pions (p. 53) cannot be regarded as solid spheres at distances less than 10^{-15} m. Relativistic (c) and quantum effects ($\hbar$) become important at such distances, which is manifest by the fact that pions transmit the nuclear force over this distance. *This is the scale of size of light nuclei.* The latter length involves the electron mass and the strength of the electromagnetic interaction. This is the distance scale at which electrons are typically held by electromagnetic attraction for the heavy nucleus. *This is the scale of size of the hydrogen atom.*

A big advance occurred in the middle of the 19th century with the discovery that electric and magnetic phenomena are intimately related. In 1820, Oersted discovered that a magnetic compass needle could be deflected when an electric current passed through a nearby wire; this was the first demonstration that electric currents have magnetic effects. In 1831, Faraday discovered a complementary phenomenon: thrust a magnet into the centre of a coil and an electric current spontaneously flows in the wire. This showed that magnets can induce electrical effects. Faraday's discovery eventually led to the development of electric generators and alternators, Bell's original telephone, transformers, and a variety of modern electrical techniques. It also gave crucial impetus to understanding the relation between electricity and magnetism.

In 1864, Maxwell produced his celebrated equations containing all the separate laws responsible for these various phenomena. This united electricity and magnetism into what is now called *electromagnetism*. In addition, he predicted the existence of things that had previously been unsuspected, the most notable being electromagnetic radiation and the realisation that light consists of fluctuating electric and magnetic fields (see Table 2.2).

Maxwell's equations succinctly summarised all known electric and magnetic effects. In 1928, Dirac combined Maxwell's theory, Einstein's relativity, and the newly discovered quantum mechanics, and showed that the resulting theory, 'quantum electrodynamics,' enables one to calculate the effects that arise when light interacts with matter, in particular with electrically charged subatomic particles such as the electron. Dirac's equations also predicted the existence of the "positron," an enigmatic positively charged version of the electron and the first example of what we now call "antimatter" (see Table 4.2).

When an electrically charged particle is accelerated by an electric or magnetic force, an electromagnetic wave is radiated. In quantum electrodynamics this wave behaves as if it were a series of particles, 'photons,' and so we regard the particle's acceleration as resulting in the emission of one or more photons.

It is customary to draw a diagram to represent this (these are known as "Feynman diagrams" after their inventor Richard Feynman). An electron is represented by a straight line and a photon by a wiggly line. Time runs from left to right in the diagrams in Figure 4.2 and so these represent an electron coming from the left and emitting one photon which is in turn absorbed by another electron. The photon transfers energy from the first to the second electron, which upon gaining this energy is accelerated.

Table 4.2 Antimatter

In 1928, Paul Dirac produced his equation describing the motion of electrons in the presence of electromagnetic radiation. He found that he could only do this in a manner consistent with relativity if in addition to the electron there exists a 'positron.' The positron has identical properties to the electron but with positive electrical charge in place of negative. The positron is the first example of an antiparticle, a piece of antimatter.

Antimatter is a mirror image of matter, possessing an equal and opposite sign of electrical charge but otherwise responding to natural forces much as the matter equivalent. Thus protons and antiprotons have the same mass and behave the same way, but have equal and opposite charges. Similarly, electrons and positrons have opposite charges but are otherwise alike.

Matter and antimatter can mutually annihilate and convert into radiant energy. The amount of energy (E) is given by Einstein's famous equation

$$E = mc^2$$

where c is the velocity of light and m is the total mass annihilated. Conversely, if enough energy is contained in a small region of spacetime, then matter and antimatter can be produced in equal abundance. In the laboratory this is frequently done. High-energy electrons and positrons annihilate to produce radiant energy, which in turn produces new varieties of matter and antimatter. This is a means of creating and studying varieties of matter that are not abundant on Earth. Exotic particles with properties such as "charm" and "bottom" were found this way (Chapter 9). At the start of the universe, in the heat of the Big Bang, there was such concentration of energy that matter and antimatter would have been created from it in vast, equal quantities. Yet today our universe is built from matter almost to the exclusion of antimatter: our atoms consist of electrons and protons; no examples exist of positrons encircling antiprotons other than such atoms of "antihydrogen" that are made at CERN in Geneva with the intention of comparing the physics of such atoms and antiatoms. Why this imbalance between matter and antimatter occurs in Nature is an important puzzle.

Antimatter particles are conventionally denoted by a line over the symbol for the equivalent matter particle. Thus, $\bar{p}$ denotes the antiproton and $\bar{\nu}$ the antineutrino. The positron however is traditionally denoted e^+ in contrast to e^- for the electron. The positron was discovered in cosmic rays by C.D. Anderson in 1932.

Newton's laws of motion tell us that acceleration occurs when a force is applied, thus the photon has effectively transmitted a force whose origin was the original electron some distance away. In this sense we say that photons mediate electromagnetic forces.

Feynman diagrams not only portray in a conceptually helpful way what is happening, but also have a precise mathematical meaning. By associating specific mathematical expressions with the various lines and vertices in the diagrams, it is possible to compute the electromagnetic properties of atomic particles to remarkable accuracy.

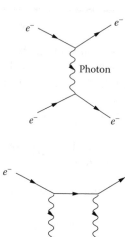

(a) An electron scatters from another electron as a result of one photon being exchanged.

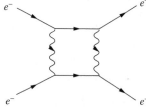

(b) After the initial scattering, a second scatter can occur as a result of a second photon being exchanged.

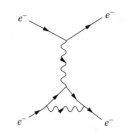

(c) In this example, one photon is transmitted to the other electron and causes scattering. The second photon is subsequently reabsorbed by the same electron that had emitted it in the first place. This contributes to the 'g-factor' or magnetic moment of the electron.

FIGURE 4.2 Feynman diagrams.

Quantum electrodynamics has been tested time and again for more than 60 years. Its predictions have been verified to ten significant figures, a testament both to the ingenuity of experimentalists and to its validity as the theory of interactions between light and charged particles. It well deserves its acronym of QED.

NUCLEAR FORCES

Whereas the attraction between opposite electric charges is the source of an electron's entrapment around the nucleus, the mutual repulsion of like charges creates a paradox in the existence of the nucleus itself: many positively charged protons are packed tightly together despite experiencing the intense disruptive force of their mutual electrical repulsion. The attractive force that holds them there must be immensely powerful to overcome this electromagnetic repulsion and, in consequence, it was named the 'strong' force. (Modern ideas imply that this results from a more fundamental force (see Chapter 7) and that "strong" is a misnomer, but I will adhere to it for historical reasons.)

Table 4.3 Yukawa and the Pion

In everyday experience, energy appears to be absloutely conserved. Quantum mechanics shows that small amounts of energy (ΔE) can be 'borrowed' for a time (Δt), where

$$\Delta E \times \Delta t = \hbar = 6.6 \times 10^{-22} \text{ MeV} \times \text{seconds}$$

This is at the root of the transmission of forces by the exchanging of particles (such as the photon in the case of electromagnetism). When a free electron emits a photon, energy is not conserved, the imbalance being determined by the amount of energy carried by the photon. The more energy the photon carries, the sooner must that energy be repaid and the less distance the photon will have travelled before being absorbed by another charged particle and energy balance restored.

A photon has no mass, so it is possible for the photon to carry no energy at all. In this case it could voyage for infinite time and so transmit the electromagnetic force over infinite distance. Contrast this with the nuclear force, which binds nucleons to one another so long as they are less than about 10^{-14} m apart, but does not act over larger distances.

This phenomenon led Hideki Yukawa to postulate that the carrier of the strong force had a mass. His reasoning was that energy and mass are related by Einstein's equation $E = mc^2$. Thus emission of the force-carrying particle would always violate energy conservation by at least mc^2 and hence the particle must be reabsorbed not later than time t:

$$t = \hbar/mc^2 \equiv \hbar/ \text{ some number of MeV}$$

Since it travels at less than the speed of light, the maximum distance it can travel and transmit the force is:

$$\text{Max. distance} = ct = \hbar/ \text{ mass in MeV}$$

Knowing that nuclear forces only occur over less than 10^{-14} m led Yukawa to propose that their carrier (the 'pion') had a mass of about 140 MeV. The subsequent discovery of the pion with this very mass led to Yukawa winning the Nobel Prize for physics in 1949.

One empirical feature of the strong force is that protons and neutrons experience it, whereas electrons do not. This suggests that protons and neutrons possess some sort of 'strong charge' which electrons do not (we shall see what this is in Chapter 7). By analogy, this led to the proposal that just as the electromagnetic force is carried by a photon, so there should be a carrier of the strong nuclear force, and the particle now called the pion was postulated.

From the observation that the nuclear force only acts over a few fermi (1 fermi = 10^{-15} m) as against the potentially infinite range of the electromagnetic force, Yukawa (Table 4.3) computed that the pion had a mass about 1/7 of a proton mass

Table 4.4a Strength of the Weak Force: I

Gravity and the electromagnetic force have infinite ranges whereas the weak and strong nuclear forces operate only over distances of about 1 fermi (10^{-15} m). Table 4.3 told how the zero mass of the photon enables it to be emitted while conserving energy and momentum. The photon is therefore free to exist forever and transmit electromagnetic forces over infinite distances. By contrast, production of a massive particle such as the pion violates energy and momentum conservation. The larger the energy account is overdrawn, the sooner it must be repaid (uncertainty principle p. 25). Here is the source of the short-range nature of the ensuing nuclear force between nucleons.

When a neutrino converts into an electron, a W^+ is emitted. With a mass of around 80 GeV, this causes a huge imbalance in energy and the amount of time that this can be tolerated is correspondingly very short. In this time it is very unlikely to travel even 1 fermi. Thus at such distances it is very unlikely that particles will experience this force as compared to the electromagnetic whose effects are easily transmitted that far.

The feeble strength of the "weak" force thus derives directly from the huge mass of its carriers — the W^+, W^-, and the Z^0 bosons — and it is this that obscures the fact that the intrinsic strength of the the W and Z coupling to the electron is comparable to that of a photon. Once this was proved in experiment (Chapter 8), the weak and electromagnetic forces were realised to be intimately connected. This is at the root of the modern theory uniting them in a single electroweak force, which combines electromagnetism and the weak force within it.

Table 4.4b Strength of the Weak Force: II

The strength of the electromagnetic force is expressed by the dimensionless quantity $\alpha \sim 1/137$. The strength of the weak force is expressed in terms of G_F, where

$$G_F \sim 10^{-5}/m_p^2; \quad (m_p = proton\ mass \sim 1\ \text{GeV})$$

known as the Fermi constant after Enrico Fermi who made the first attempt at constructing a theory of the weak force (Chapter 8). It is the smallness of 10^{-5} relative to $1/137$ that gives meaning to the concept of "weak" force, but this is actually misleading. First, note that G_F has dimensions, with the result that

$$G_F/\alpha \sim 10^{-3}/\text{GeV}^2$$

So in nucleon beta-decay and manifestations of the weak force at energy scales of some 1 GeV, the ratio is about 10^{-3} and it is truly weak. But at higher energies, the *dimensionless* measure

$$G_F/\alpha \times (\text{Energy of experiment in GeV})^2$$

can be of order unity. More precisely, a comparison that is relevant is:

$$G_F/\alpha \times m_W^2$$

where $m_W \sim 80$ GeV is the mass of the carrier of the force (the analogue of the photon in electromagnetic forces). This huge mass of the W causes the ratio to be on the order of unity; the basic strengths of the electromagnetic and weak forces are essentially the same, the latter only appears weak at low energies due to the huge mass of its carrier, the W. This was verified to be the case in experiments at CERN in the 1990s (more details appear in Chapters 8 and 10).

(as against the masslessness of the photon). The eventual discovery of the pion in 1947 with this mass confirmed it as the carrier of the strong nuclear force over distances of the order of a fermi.

In nuclear decay involving α or γ emission, the strong and electromagnetic forces are at work. In both these types of decay, the number of neutrons and protons is separately conserved. While these two forces control these and almost all other nuclear phenomena so far observed, there is one remaining process that they cannot describe. This is the source of the last of Becquerel's three radiations: the β-decay process.

Emission of beta particles (electrons) occurs when a neutron in the nucleus becomes a proton. The net electrical charge is preserved by the emission of the electron

$$n^0 \rightarrow p^+ + e^- + \bar{v}^0$$

but the number of neutrons, protons, and electrons changes. The electromagnetic and strong forces do not have such an ability: the number of neutrons and protons or electrons is conserved when these forces act. The agent responsible for the neutron's β-decay is known as the 'weak' force, being some hundred thousand times less powerful than the strong nuclear force when acting in this way (see Figures 4.3a, b).

An important property of β-decay is the fact that it produces neutrinos (v^0) (technically antineutrinos $\bar{v}^0$, Table 4.2). These are electrically neutral and thus inert to electromagnetic forces. Furthermore, like the electron, they are blind to the strong force. The weak force is the only one that measurably affects them. This makes neutrinos a unique tool for studying the weak force; by firing them at targets and studying how their flight is disturbed, we are seeing directly the weak force at work (see Figure 4.4). (We meet the neutrino in more detail in Chapter 12.)

When neutrinos are fired at matter they are most noticeably converted into electrons but examples of the action of the weak force also occur where neutrinos scatter without changing their identity. There are similarities between these processes and the familiar electromagnetic scattering of electrons from charged matter.

Just as the photon is the carrier of the electromagnetic force and the pion the carrier of the strong force across the atomic nucleus, so is there a carrier of the weak force. When neutrinos are converted into electrons, this is due to the action of an electrically charged force carrier, the 'W boson.' When neutrinos scatter and preserve their identity, it is the neutral 'Z boson' that is responsible. The W and Z are nearly 100 times more massive than a proton, which was too large for them to be

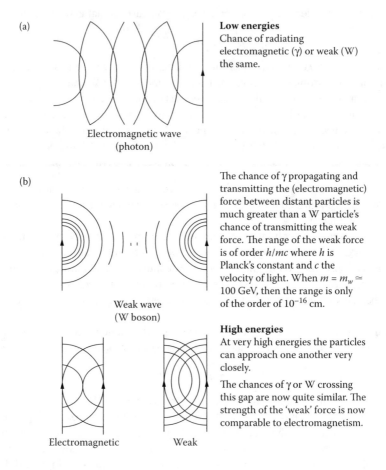

(a)

Electromagnetic wave
(photon)

Low energies
Chance of radiating
electromagnetic (γ) or weak (W)
the same.

(b)

Weak wave
(W boson)

The chance of γ propagating and
transmitting the (electromagnetic)
force between distant particles is
much greater than a W particle's
chance of transmitting the weak
force. The range of the weak force
is of order h/mc where h is
Planck's constant and c the
velocity of light. When $m = m_w \simeq$
100 GeV, then the range is only
of the order of 10^{-16} cm.

High energies
At very high energies the particles
can approach one another very
closely.

The chances of γ or W crossing
this gap are now quite similar. The
strength of the 'weak' force is now
comparable to electromagnetism.

Electromagnetic   Weak

FIGURE 4.3 W bosons and the weak force.

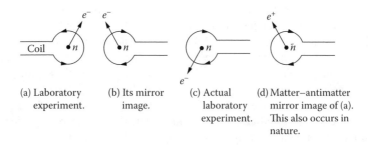

(a) Laboratory
experiment.

(b) Its mirror
image.

(c) Actual
laboratory
experiment.

(d) Matter–antimatter
mirror image of (a).
This also occurs in
nature.

FIGURE 4.4 Parity: mirror symmetry.

If mirror symmetry were an exact property of Nature, it would be impossible to tell whether a film of an experiment has been made directly or by filming the view in a mirror in which the experiment has been reflected. This is equivalent to saying that Nature does not distinguish between left and right in an absolute way. This is the case for phenomena controlled by gravity, the electromagnetic or strong forces. As these control most observed phenomena, it had been assumed that left-right symmetry is an inherent property of all subatomic processes. But in 1956 mirror symmetry was discovered to be broken in weak interactions.

The historic experiment involved the beta-decay of cobalt nuclei, but we can illustrate it for the beta-decay of a single neutron. An electric coil makes a magnetic field that interacts with the neutron's magnetic moment and aligns its spin along the direction of the field's axis. The electrons produced in the decay are preferentially emitted upwards (a). Viewed in a mirror, the electrons are emitted upwards if an electric current flows in the coil in the *opposite* direction (b). If mirror symmetry is a property of nature, the electrons should still be emitted upwards in the lab when the current flows in the opposite direction. However, what is observed is that the electrons are emitted downwards (c). More precisely, the electrons are emitted on that side of the coil from which the current is seen to flow clockwise. By this violation of mirror symmetry, nature provides an absolute meaning to left and right. If we imagined a magic mirror that also interchanged matter and antimatter, then the combined exchange of left-right and matter-antimatter would restore the symmetry in this process (d).

produced in accelerators until the 1980s. It was following the discovery of W and Z, which apart from their huge masses appear to be similar to the photon, that we now recognise the electromagnetic and weak forces to be two manifestations of a single 'electroweak' force (Chapter 8).

The history of the weak interaction theory and more about the W and Z bosons are deferred to Chapter 8, after I have described the search for the carrier of the strong force: the pion. The discovery of the pion (in 1947) opens up a Pandora's box that would confuse physicists for nearly two decades. When the dust settles our story will have reached the 1960s. A deeper layer of matter consisting of 'quarks' will be perceived inside nuclear matter. Not until then will we be able to contemplate a complete theory of the weak force, and also understand the origin of the strong force.

All four forces are needed for life to emerge. To recap: the strong force compacts atomic nuclei; the weak force helps transmute the elements within stars to build up the richness of the periodic table; the electromagnetic force ensnares electrons and builds atoms and molecules; while gravity enabled the star to exist in the first place. It is remarkable that just these four forces are needed for us to be here. The strengths and characters of these forces are tightly related. For example, had the weak force been just three times more powerful than in reality, the life-giving warmth of the sun would have ended long ago; alternatively, had the force been more feeble, it is probable that the elements of life would not yet have been "cooked" within the stars.

5 Nuclear Particles and the Eightfold Way

PIONS

Protons and neutrons ("nucleons") are gripped in the atomic nucleus by a strong force (Figure 4.1). The electromagnetic force is transmitted by photons, and, in 1935, Hideki Yukawa proposed that the strong force also has an agent — the pi-meson, or pion, labelled π. In his theory, it is the exchange of a pion between pairs of protons or neutrons that attracts them (Figure 5.1). However, unlike the electromagnetic force which has an infinite reach, the influence of the strong force extends hardly beyond the breadth of two nucleons, a mere 10^{-15} m. To explain this, Yukawa proposed that the pion had a mass. Had the pion been massless like the photon, the strong force would have been infinite range like the electromagnetic force. However, the range of 10^{-15} m required the pion to have a mass of about 1/7 that of a proton.

The neutron has emitted a pion and remained a neutron in Figure 5.1a and the proton emitted a pion and stayed a proton in Figure 5.1b. In each example the pion has no electrical charge (the total charge is always conserved; the neutron and proton gave none up, so the pion carries none). To denote its electrical neutrality we label it π^0.

A pion can be emitted by a neutron and absorbed by a proton, or vice versa Figure 5.1c whereby the neutron and proton exert a force on one another. Yukawa's theory of the nuclear force also required that the neutron and proton can exchange their positions (d) or (e). There is still a neutron and a proton at the start and at the finish as in Figure 5.1c, but this time electrical charge has been carried across. If the forces between neutrons and protons are transmitted by pions, then three varieties of pion are called for. These are denoted π^+, π^-, π^0 depending upon whether they carry positive, negative, or no electrical charge, respectively, and their masses are almost identical. Pions are unstable, surviving for less than 10^{-8} seconds before decaying into photons, neutrinos, and electrons or positrons.

When Yukawa proposed the existence of the pion, the best hope of finding it lay in the cosmic ray showers that continuously bombard the Earth. Cosmic rays consist of heavy nuclei, protons, electrons, and similar objects produced in stars and accelerated to extreme energies by magnetic fields in space. By studying them in detail, it was hoped to discover other types of matter that had not previously been observed on Earth.

Originally, cosmic rays were detected using a cloud chamber, a device devised by C.T.R. Wilson in 1911 (see also Table 5.1). When an electrically charged particle (such as a cosmic ray) passes through supersaturated mist, it ionises the gas's atoms. Droplets of water settle on these ions, forming a vapour trail (similar to that from a highflying aircraft) which reveals the particle's trajectory.

The nature of the particle can be deduced from the form of the trail. Massive particles (such as atomic nuclei) plough straight through and cause many drops to form, yielding a thick straight track. Electrons are so light that collisions with the

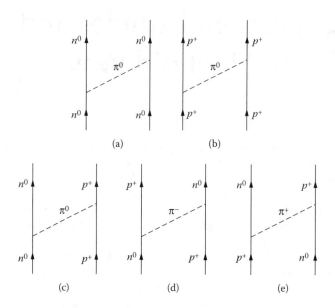

FIGURE 5.1 Exchange of a pion between neutrons and protons.

atoms in the chamber easily divert them from their path, and a rather diffuse wavy track results.

Put a cloud chamber in a strong magnetic field and the trajectories of charged particles will curve. The steepness of the curve reveals the momentum of the particles: the lower the momentum, the sharper the curve, and the direction of the deflection shows whether the charge on the particle is positive or negative. The more massive particles tend to leave denser trails. Thus a lot of information about charged particles can be easily obtained. Uncharged particles, on the other hand, leave no track and their presence can only be detected indirectly (compare the invisible man, p. 33).

What will one of the pions look like? Being heavier than an electron but lighter than a proton, a pion in the cosmic rays tends to have momentum intermediate between them. On passing through a magnetic field, the direction of the bending will be the same as that of an electron in the case of π^-, but the same as that of a proton in the case of a π^+ (the π^- and π^+ bending in opposite directions).

In 1936, C.D. Anderson and S.H. Neddermeyer found such a track in a cloud chamber. The particle responsible had a positive charge and a mass that was slightly lighter than Yukawa's prediction. A negative version of the particle was found by J. Street and E. Stevenson at about the same time. However, no evidence for an electrically neutral partner turned up. This was puzzling. More disturbing was that the particle showed no desire to interact with nuclei. As the *raison d'etre* for Yukawa's particle had been that it provided the grip that held the nucleus together, then it must necessarily have a strong affinity for nuclear material.

The resolution of the puzzle was that this was *not* Yukawa's particle, nor did it have any role as a carrier of the strong nuclear force.

Table 5.1 The Bubble Chamber, and Beyond

Donald Glaser was gazing at a glass of beer, watching the bubbles rise to the surface and musing about the minute imperfections of the glass container on which the bubbles form. From this contemplation the idea of the bubble chamber was born.

In a bubble chamber, the paths of charged particles are made visible. The chamber contains liquid that is on the point of boiling and a piston in the chamber suddenly lowers the pressure in the liquid causing the boiling to start. Bubbles of gas start to grow if there is some central stimulus such as the irregularities on the beer glass. If a charged particle passes through the liquid at the critical moment, it ionises some of the liquid's atoms, which act as centres for bubble formation. The trail of bubbles reveals its trajectory. Sometimes the particle hits one of the nucleons in the liquid and produces new particles, which show up as multiple tracks emanating from the point of collision.

Today bubble chambers have been superceded by other devices. Spark chambers consist of parallel sheets of metal separated by a few millimetres and immersed in an inert gas such as neon. A charged particle leaves an ionised trail in the gas and, if high voltage is applied to the sheets, sparks will form along

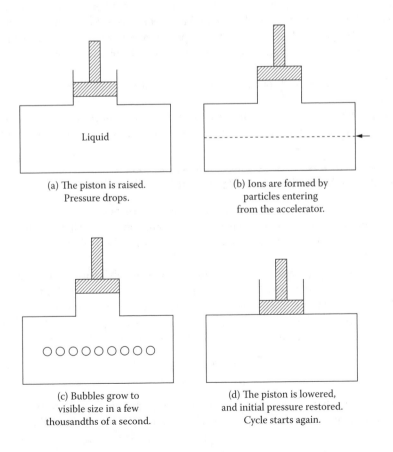

(a) The piston is raised.
Pressure drops.

(b) Ions are formed by
particles entering
from the accelerator.

(c) Bubbles grow to
visible size in a few
thousandths of a second.

(d) The piston is lowered,
and initial pressure restored.
Cycle starts again.

the ionised trails. Magnetic fields steer the particles left or right depending on the sign of their charge. The spark chamber can operate up to a thousand times faster than a bubble chamber. In modern experiments, a range of sophisticated electronic devices are used to detect particles. A fuller description appears in Frank Close's *Particle Physics: A Very Short Introduction.*

We now know that what Anderson discovered in 1936 was a muon (denoted μ), a particle that is similar in many ways to an electron except that it is about 200 times heavier. This was quite unexpected and the reasons for its existence remained a total mystery for the next 40 years.

Yukawa's particle, the pion, was finally discovered in 1947 by C.F. Powell. He had suspected that interactions of pions in the atmosphere might be preventing most of them from ever reaching the Earth, and so he set up an experiment high on the Pic du Midi in the French Pyrenees. Instead of a cloud chamber, Powell used emuulsions, similar to those used in photogrpahy. In the late 1940s, the development of special photographic emulsions, which could easily be carried aloft by balloons, brought physicists their first beautiful images of the interactions of high-altitude cosmic rays. These emulsions were especially sensitive to high-energy particles; just as intense light darkens photographic plates, so can the passage of charged particles. We can detect the path of a single particle by the line of dark specks that it forms on the developed emulsion. The particle literally takes its own photograph. Cosmic rays passing through affected the chemical and produced a dark track when the photographic emulsion was developed.

So it was that Powell found the pion, with a mass of 140 MeV, as Yukawa had predicted.

When the π^+ or π^- decay, they usually produce a μ^+ or μ^- and a neutrino. See Table 5.2. Ironically, the muon detected by Anderson in 1936 was the progeny of the

Table 5.2 More than One Variety of Neutrino

There are three different varieties, or "flavours" of neutrino, known as the electron-neutrino ν_e, muon-neutrino ν_μ, and tau-neutrino ν_τ. They are siblings of the charged leptons whose names they share. The origin of this discovery is as follows.

When a π^+ decays, it usually produces a positively charged muon and a neutrino. Less often it will produce a positron (antielectron) and a neutrino. These two neutrinos are distinct, denoted ν_μ and ν_e, to denote that they were born in conjunction with a μ and e, respectively. When these neutrinos, hit neutrons, they can convert it to a proton plus an electron or muon:

$$\nu_e + n \rightarrow p + e^-$$
$$\nu_\mu + n \rightarrow p + \mu^-$$

The neutrino born with an electron (muon) always produces electrons (muons). Somehow the neutrino carries a memory of how it was born in the weak interaction. A third variety of neutrino, born in the decay of the τ is the ν_τ; when it hits nuclear targets, it gives rise to the τ, analogous to the above.

Recently these neutrinos have been found to change their flavour very subtly while in flight. This arises because they have (very small) masses and can "oscillate" back and forth from one variety to another en route (Chapter 12). Chapter 10 describes how we know that there are no more than three neutrinos such as these.

---

particle that he had been seeking. The π^+ and π^- were finally produced in particle accelerators at Berkeley in 1947 as products from the collision of alpha particles and carbon nuclei. The π^0 was subsequently found in 1949 as a product of similar collisions.

So apart from the unexpected appearance of the muon, everything was turning out rather well. Table 5.3 summarises the situation in 1947.

STRANGE PARTICLES

As studies of cosmic rays continued, so further particles were discovered. These were debris produced from collisions between the high-energy pions and the nuclei of atoms. In that brief moment of interaction it was the strong force that was at work, bringing the pions into close contact with the protons and neutrons, and transforming the kinetic energy of the pion's motion into new material particles: $E = mc^2$ in action.

Having been so strongly produced this way, one would have expected these new particles to have decayed rapidly back into pions and protons, the very particles which had been responsible for their production. However, this did not happen. Instead of

Table 5.3 Particle Summary 1947

Name/symbol		Charge	Mass (proton as unit)	Stable	Does it feel the strong force?
Electron (1897)	e	-1	$\frac{1}{1800}$	yes	no
Neutrino (1931–56)	ν	0	0(?)	yes	no
Muon (1936)	μ	-1	$\frac{1}{2}$	$\tau \sim 10^{-6}$ s*	no
Proton	p	$+1$	1	yes (?)	yes
Neutron (1932)	n	0	1	$\tau \sim 15$ min when free	yes
Pion (1947)	$\pi^\pm$	± 1	$\frac{1}{7}$	$\tau \sim 10^{-8}$ s	yes
[(1949)	π^0	0	$\frac{1}{7}$	$\tau \sim 10^{-16}$ s	yes]

* The symbol τ means lifetime.

The π^+, π^-, and π^0 had been predicted by Yukawa. Following the $\pi^\pm$ discovery in 1947, the observation of the uncharged sibling, π^0 was confidently awaited. I include it is in this list for completeness although it was not directly detected until 1950. Similar comments apply to the neutrino — accepted by the physics community but not detected until 1956 (Chapter 3). Two varieties of antimatter had also been detected: the antielectron (positron) e^+ discovered in 1932 and the antimuon, μ^+, discovered in 1936.

decaying into a pion or proton in a mere 10^{-23} seconds, as should have been the case if the strong force was at work, they lived up to 10^{-10} seconds (a million billion times longer than expected). Once they were produced, the effects of the strong force seemed to have been cut off. To illustrate by how much its effects are postponed, one scientist said 'It was as if Cleopatra fell off her barge in 40 BC and hasn't hit the water yet.' Another peculiar property was that these novel particles were always produced in pairs. These unusual properties in production and decay caused them to become known as 'strange' particles.

Among the strange particles is a quartet somewhat heavier than pions. These are the electrically charged 'kaons,' K^+ and K^-, and two varieties of neutral kaon denoted K^0 and $\bar{K}^0$, all with masses of about 500 MeV. The discovery of the K^0 is usually attributed to Rochester and Butler who found it in cosmic rays in 1947. The uncharged K^0 left no track of its own but decayed into two charged particles (now known to be π^+ and π^-) and so left a distinctive V-shaped vertex in the photograph. This caused them initially to be referred to as V particles. It subsequently turned out that K mesons had been seen, but not recognised, in cloud chamber photographs of cosmic rays studied by Leprince-Ringuet and Lheritier as early as 1944.

Cloud chambers were superseded by bubble chambers. Instead of trails of water drops in a cloud, the tracks in a bubble chamber consisted of bubbles formed in a superheated liquid (see Table 5.1). The density and changes in directions of the tracks in a bubble chamber gave information on the particles' properties analogous to those in a cloud chamber. However, the bubble chamber had several advantages, in particular that it was both target and detector. Collisions between the projectile particles and the nuclei in the bubble chamber liquid can be directly observed, and not only the original particle but also the nuclear fragments can be seen. (For an extensive collection of bubble chamber images, see *The Particle Odyssey* in the bibliogrpahy.) The art of the experimentalist was to study the resulting tracks and determine what particles caused them. By such means the new strange world began to be interpreted.

(Today bubble chambers are no longer used. Electronic devices record the particles with precision far beyond what was possible with bubble chambers. Electronics also enables forms of experiment to be performed that would be impossible for bubble chambers; we learn more about these in Chapters 10 and 11. Emulsions are still used in specialist experiments but in association with electronic aids.[1])

In any reaction initiated by pions or nucleons, the strange kaons are always produced in partnership with other strange particles such as the Lambda (Λ), Sigma (Σ), or Xi (Ξ). These particles are more massive than the proton and neutron, and from their properties we now recognise them as strange baryons (Table 5.4), with baryon number $+1$ the same as the proton. The striking feature of the pair production can be illustrated as follows.

If a negatively charged pion hits a proton, then baryon number (B) and charge (Q) conservation (Table 5.5) would allow both

$$\pi^- + p \rightarrow K^- + \Sigma^+ (B_{total} = 1; Q_{total} = 0)$$

---

[1] For more about detectors, read *Particle Physics: A Very Short Introduction*, by Frank Close.

and

$$\pi^- + p \rightarrow \pi^- + \Sigma^+ (B_{total} = 1; Q_{total} = 0)$$

to occur. Indeed, it uses less energy to produce light pions than the heavier kaons, so the second reaction should be more copious than the first. The puzzle was that this reaction has never been seen, in contrast to the former which has been seen billions of times.

To rationalise this, in 1953, Gell-Mann, Nakano, and Nishijima proposed that there exists a new property of matter, which they named 'strangeness,' and that this strangeness is conserved in strong interactions (Table 5.4). A pion or proton has no strangeness. When they interact and produce a particle with strangeness +1, another particle with strangeness −1 must also be produced so that the total amount of strangeness is conserved. Thus if we arbitrarily assign strangeness value +1 to the K^+, we can deduce the strangeness of all other particles by finding out which reactions happen and which do not. For example, the Sigma-minus (Σ^-) has strangeness −1:

$$\pi^0 + n \rightarrow K^+ + \Sigma^-$$

Strangeness : $0 + 0 \rightarrow (+1) + (-1)$ (net strangeness zero)

and the unobserved reaction is forbidden because strangeness would not be conserved.

$$\pi^- + p \nrightarrow \pi^- + \Sigma^+$$

Strangeness : $0 + 0 \rightarrow 0 + (-1)$

This scheme can be applied to all strange particles and a totally consistent picture emerges. The results are as follows: the Λ and Σ have strangeness −1, the Ξ has strangeness −2, the K^+ and K^0 have strangeness +1, while the K^- and $\bar{K}^0$ have strangeness −1.

At this point you may well ask 'But what is strangeness?' It is a property of matter, analogous to electric charge, which some particles have and others do not. This may sound rather like an arid answer, but it is important to realise that physicists invent concepts and rules to enable them to predict the outcome of natural processes. By inventing strangeness we can successfully predict which reactions will or will not occur. Although the deep question of what strangeness 'is' is currently metaphysics, insights have been gained into the reason why the various particles carry the particular magnitudes of strangeness that they do (Chapter 6).

One further important property of strangeness concerns its role when hadrons decay. The Σ^0 baryon is more massive than the Λ^0 and so can lose energy by radiating a photon (γ) and converting into a Λ^0 while conserving strangeness

$$\Sigma^0 \rightarrow \Lambda^0 + \gamma$$

which the Σ^0 does within 10^{-20} seconds of its birth. But the Λ^0 is the lightest strange baryon; it cannot decay into lighter particles if baryon number and strangeness are both to be conserved. Thus we would expect the Λ^0 to be absolutely stable.

Table 5.4 Hadrons and Leptons

As the number of particles proliferated, attempts were made to classify them into families with similar properties. Some, such as the electron and neutrino, do not feel the strong force and are called leptons. (The name is taken from a small Greek coin.) Particles that feel the strong interactions are named hadrons. The hadrons are divided into two categories: mesons (such as the pion) and baryons (such as proton).

Hadrons carry an intrinsic angular momentum or 'spin,' which is a multiple of Planck's constant $\hbar$ in magnitude. For baryons, this is half-integer: $\frac{1}{2}, \frac{3}{2}, \frac{5}{2} \ldots$, whereas for mesons it is an integer: $0, 1, 2 \ldots$. All leptons so far discovered have spin $\frac{1}{2}\hbar$.

Table 5.5 Charge and Baryon Number Conservation in Particle Interactions

Charge

Charge is expressed in units of the proton's charge. All particles so far detected have charges that are integer multiples of this quantity. The total charge never changes in any reaction or decay, for example in

$$\pi^- + p \rightarrow \pi^0 + n^0$$

charge (Q): $-1 + +1 \rightarrow 0 + 0$

the totality is preserved at zero. Thus charge conservation forbids

$$\pi^- + p \rightarrow \pi^+ + p$$

which has indeed never been seen.

Baryon number

Electrical charge conservation does not prevent

$$p^+ \rightarrow e^+ + \pi^0$$

but this decay of the proton has not been seen: the proton is very stable with a half-life that is at least twenty orders of magnitude greater than the life of the present universe! It is possible that protons are absolutely stable. This has been rationalised by inventing the principle of baryon conservation.

The baryon number of the proton and neutron is defined as $B = 1$; that of the lepton or photon is defined to be zero. Mesons, such as π, also have zero baryon number. Thus

$$p \rightarrow e^+ + \pi^0$$
$$B : 1 \quad 0 \quad 0$$

is forbidden, because the baryon number changes. (Recently there have been theoretical suggestions that baryon number might not be conserved after all, but there is no evidence to support this yet.)

Strange particles also are assigned baryon number. Particles such as Σ, Λ, Ξ which decay and ultimately leave a proton in their decay products have $B = 1$. The K and other strange mesons decay into pions, photons, and leptons: they are assigned $B = 0$. Baryon conservation applies to them as well as to nonstrange particles.

And so it would be if the strong and electromagnetic forces were the only ones in Nature. In fact, it is metastable. After about 10^{-10} seconds, the Λ^0 decays and in doing so, the amount of strangeness changes. As an example, one of its decay modes is

$$\Lambda^0 \to p^+ + e^- + \bar{\nu}^0$$

the initial particle having strangeness -1, the final state having none. Similarly, the K^-, for example, decays and violates strangeness as follows:

$$K^- \to \mu^- + \bar{\nu}^0$$

These decays have similar behaviour to those of the neutron and the lightest mesons:

$$n^0 \to p^+ + e^- + \bar{\nu}^0$$

$$\pi^- \to \mu^- + \bar{\nu}^0$$

which are well-known manifestations of *weak interactions*. Thus it appears that strangeness is conserved in strong interactions and violated in weak interactions (Figure 5.2 and Table 5.6a).

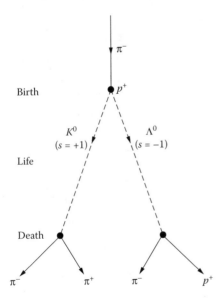

A pion incident from a cosmic ray or particle accelerator hits a proton in the atmosphere or a bubble chamber.
Two particles are produced. The reaction is written:

$$\pi^- + p \to K^0 + \Lambda^0$$

Strangeness

$$0 + 0 \to (+1) + (-1)$$

Strangeness is balanced by production of a pair of strange particles.

Strangeness likes to be conserved. This prevents the K^0 or Λ^0 dying.

Eventually weak interaction steps in. Strangeness is no longer conserved.

Electrical charge is always conserved. The neutral K^0 and Λ^0 each produce a pair of charged particles. These show up in photographs as a characteristic V.

The Λ baryon produces a proton in its decay; the K meson decays to π mesons: baryon number is conserved.

FIGURE 5.2 The strange life.

Table 5.6a Known and Postulated Hadrons 1953

Name/symbol	Charge	Mass (proton as unit)	Strangeness	Stable
Proton p	+1	1	0	Yes (?)
Neutron n	0	1	0	$\tau \sim 15$ min free
Pion $\pi^{\pm}$	± 1	$\frac{1}{7}$	0	10^{-8} s
π^0	0	$\frac{1}{7}$	0	10^{-16} s
Kaon $K^{\pm}$	± 1	$\frac{1}{2}$	± 1	10^{-8} s
$K^0, \overline{K^0}$	0	$\frac{1}{2}$	± 1	10^{-8} or 10^{-10} s
Sigma $\Sigma^{\pm}$	± 1	1.2	-1	10^{-10} s
[(1958) Σ^0	0	1.2	-1	10^{-20} s]
Lambda Λ^0	0	1.1	-1	10^{-10} s
Xi [(1959) Ξ^0	0	1.3	-2	10^{-10} s]
Ξ^-	-1	1.3	-2	10^{-10} s

The Λ^0 was too light to be mistaken for the uncharged sibling of the Σ and Σ^-: a Σ^0 was predicted. The Gell-Mann and Nishijima strangeness scheme required that an uncharged Ξ partnered the observed Ξ^-. The Σ^0 and Ξ^0 were observed in 1958 and 1959, respectively. There was much confusion in understanding the K^0, $\overline{K^0}$ mesons which was not resolved until 1956. Baryons are shown in ordinary and mesons in bold type.

Table 5.6b Hadron Summary c. 1960

Name/symbol	Charge	Mass (proton as unit)	Strangeness	Stable
Proton p	+1	1	0	Yes (?)
Neutron n	0	1	0	$\tau \sim 15$ min free
Pion $\pi^{\pm}$	± 1	$\frac{1}{7}$	0	10^{-8} s
π^0	0	$\frac{1}{7}$	0	10^{-16} s
(1961) **Eta** η^0	0	$\frac{1}{2}$	0	10^{-19} s
Kaon $K^{\pm}$	+1	$\frac{1}{2}$	± 1	10^{-8} s
$K^0, \overline{K^0}$	0	$\frac{1}{2}$	± 1	10^{-8} or 10^{-10} s
Sigma (1958) $\Sigma^{\pm}$	± 1	1.2	-1	10^{-10} s
Σ^0	0	1.2	-1	10^{-20} s
Lambda Λ^0	0	1.1	-1	10^{-10} s
Xi (1959) Ξ^0	0	1.3	-2	10^{-10} s
Ξ^-	-1	1.3	-2	10^{-10} s
Delta Δ^{++}	+2	1.2	0	10^{-23} s
Δ^+	+1	1.2	0	10^{-23} s
Δ^0	0	1.2	0	10^{-23} s
Δ^-	-1	1.2	0	10^{-23} s
(1961) Sigma-star				
$\Sigma^{\pm *}$	± 1	1.4	-1	10^{-23} s
$\Sigma^{0 *}$	0	1.4	-1	10^{-23} s
(1962) Xi-star				
$\Xi^{0 *}$	0	1.5	-2	10^{-23} s
$\Xi^{- *}$	-1	1.5	-2	10^{-23} s

The discoveries of Σ^0 and Ξ^0 completed an octet of baryons. The eta meson discovery in 1961 showed that mesons formed a family of eight analogous to the eight baryons. The ingredients of Yuval Ne'eman and Murray Gell-Mann's 'Eightfold way' theory (1961) was then to hand. This theory predicted that a family of ten baryons should exist. The Σ^* and Ξ^*, announced in 1962, led to Gell-Mann's dramatic prediction of the Ω^- particle (Figure 5.4) and the verification of the theory. Mesons are in bold type.

MORE HADRONS

During the 1950s and 1960s, the advent of high-energy accelerators (Table 5.7) enabled beams of protons, pions, and even kaons to be fired at nuclear targets. The debris that emerged from these collisions contained new varieties of matter that survived for only 10^{-23} seconds, the time that it takes for light to cross a proton. For example, there were the Δ (Delta) baryons, Δ^-, Δ^0, Δ^+, Δ^{++} which produce protons in their decays, such as $\Delta^+ \to p\pi^0$. These Delta particles have no strangeness. Their masses are about 1235 MeV, 30% greater than the proton.

Σ^* particles were observed with similar rapid decays into Σ and π. The properties of the decays showed them to be very similar to that of Δ into p and π. Strangeness conservation in the rapid decays shows that Σ^* particles have strangeness -1. Their masses are about 1385 MeV.

A pair of particles with strangeness -2 was found, Ξ^{*-} and Ξ^{*0}, which decayed rapidly into Ξ^- and Ξ^0. The Ξ^* masses are about 1530 MeV.

These particles, which feel the strong interactions, are generically known as 'hadrons.' (Contrast this with those particles such as the electron, muon, and neutrino, which do not respond to strong interactions, and are collectively known as 'leptons'.)

Table 5.7 Modern Particle Accelerators

Early examples of particle accelerators were described in Table 3.1. By 1945, a large machine known as a 'synchrocyclotron' existed, capable of accelerating protons to 720 MeV energy. Smashing them into nuclear targets produced the π^0 in the laboratory. The solid magnet of that machine was superceded by magnets surrounding an evacuated tube along which the protons can be accelerated, known as 'synchrotrons.'

By 1953, synchrotrons at Brookhaven, New York, and Berkeley, California, were able to accelerate protons to over 1 GeV. These produced the strange particles in the laboratory — previously one had to study them in the uncontrollable cosmic rays. The advent of these machines led to the discovery of the baryon and meson octets. In 1953, the first synchrotrons capable of 30 GeV were built at Brookhaven and at CERN, Geneva; a whole spectrum of hadrons with masses up to three times that of the proton emerged. At Fermilab near Chicago and also at CERN from the 1970s, protons were accelerated to hundreds of GeV in rings whose diameters were over a mile. Today beams of protons and antiprotons (such as at Fermilab and CERN's Large Hadron Collider); electrons and positrons (in the 1990s at LEP, CERN, and in dedicated lower-energy machines described in Chapter 11), or even electrons and protons (at Hamburg) are brought into head-on collision. Such 'colliders' are the frontier accelerators for the 21st century.

In addition to these, it is possible to fire protons at targets and thereby to create secondary beams of particles such as neutrinos. These are being used as dedicated probes of the weak force and also to investigate the nature of the enigmatic neutrinos themselves.

The apparent simplicity and order that had existed in 1935, when it was thought that only a handful of elementary particles existed, had been replaced by a new complexity; see Table 5.6b. But then a pattern of regularity was noticed in the properties of this rapidly growing 'zoo' of particles and a new simplification emerged. What Mendeleev had done in 1869 for the atomic elements, so Murray Gell-Mann and Yuval Ne'eman did for the hadrons nearly a century later, in 1960–1961.

THE EIGHTFOLD WAY: 'A PERIODIC TABLE FOR THE NUCLEAR PARTICLES'

To illustrate what Gell-Mann and Ne'eman achieved, we will begin with the mesons (pions and K particles). Mesons have electrical charges of 0, +1, or −1. They also have strangeness of 0 (the pions) or +1 (the K^+ and K^0) or −1 (the K^- or $\bar{K}^0$). We could draw a diagram with the amount of strangeness on the vertical axis and the amount of charge along the horizontal (for historical reasons the charge axis is at an angle — Figure 5.3). We now place the mesons at various points on this figure.

The K^+ has strangeness +1 and charge +1. This is the point at the top right-hand corner (where the line for charge +1 intersects the horizontal line for strangeness +1). So we place the K^+ at this point of the figure. The place that the line for positive charge intersects the line for zero strangeness is at the far right of the figure. The particle with no strangeness and with positive charge is the π^+ and so we put the π^+ there.

Continuing in this way we find a position for each and every particle and the resulting pattern is shown in Figure 5.3c. The pattern is a hexagon with a particle (π^0) at the centre.

Now we can play the same game for the baryons (the neutron, proton, Λ, Σ, and Ξ) but add one unit to the strangeness axis. The same hexagonal structure emerges when we place the particles on the figure but this time we find two particles at the centre instead of the one in the previous example (Figure 5.4).

This similarity in the patterns is very striking. To make them identical would require an eighth meson without strangeness or charge so that it could accompany the π^0 in the centre spot. The discovery of the eta meson (η) in 1961, mass 550 MeV, with no charge or strangeness, completed the pattern.

The common pattern for the baryons and for the mesons suggests some important underlying relation between them. These patterns of eights were named the 'Eightfold Way' by Gell-Mann.

We can play the same game with the Δ, Σ^*, and Ξ^* particles because these particles seem to form a family: each decays into analogous members of the octet that contained the proton (Figure 5.4a). Now put these particles onto a similar figure. Instead of a simple hexagon, this time we find a hexagon with extra particles at the top corners (the Δ^0 and Δ^{++}, respectively).

The theory that Gell-Mann and Ne'eman had developed of the Eightfold Way led them to expect that a group of ten should exist. (In Chapter 6 we see how these patterns emerge due to an underlying 'quark' structure in nuclear matter and all hadrons). The pattern in Figure 5.4b should be completed by extending the pattern

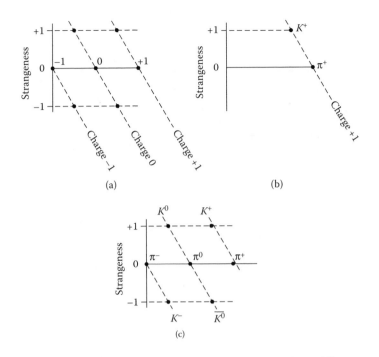

Hadrons—particles that interact through the strong nuclear force—fall into patterns according to properties such as electric charge (plotted horizontally) and strangeness (plotted vertically). These patterns are called the 'Eightfold Way.' This is illustrated in this and Figure 5.4.

(a) Make two axes: strangeness on the vertical and electrical charge on the horizontal slanted as shown.
(b) Particles with charge +1 will lie on the charge +1 line. The K^+ has strangeness +1 and so occurs at the point common to strangeness +1 and charge +1. The π^+ has no strangeness.
(c) The K and π particles' positions on the figure yield a hexagonal pattern with a π^0 at the centre.

FIGURE 5.3 The Eightfold Way.

at the bottom, thereby forming an inverted triangle. The position of the particle that would complete the pattern (which has ten members and is called a decuplet) would occupy the position indicated by * in the figure. It would have strangeness −3 and have negative charge. Gell-Mann named it the Omega Minus (Ω^-), "Minus" referring to its negative electrical charge and "Omega" (the final letter of the Greek alphabet) in honour of it being the last particle in the pattern, and thus the final step in proving the validity of the scheme.

Furthermore, Gell-Mann was able to predict its mass. The Δ particles have zero strangeness and mass 1235 MeV. The Σ^* have strangeness −1 and mass about 1385 MeV, and the Ξ^* with strangeness −2 have masses of 1530 MeV. Each time you go down the pattern from strangeness 0 to −1 and then to −2 the mass increases by

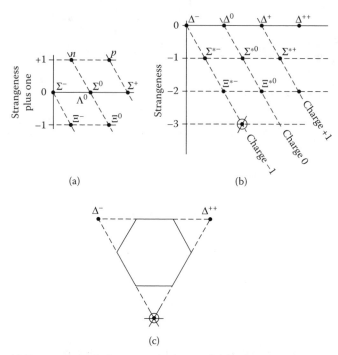

(a) The pattern for the baryons is the same as that for the mesons but
    for the presence of a second particle at the centre (Λ^0 and Σ^0). The
    discovery of η^0 at the centre of the meson pattern completed the
    correspondence.
(b) Heavier baryons were found which were related to the proton, Σ
    and Ξ baryons. The pattern is the familiar hexagon with extra
    particles at the top corners. The mathematical theory that was
    developed to describe these patterns required an inverted triangle to
    exist containing ten particles.
(c) The * denotes the position of the particle (Ω^-) required to complete
    the pattern.

FIGURE 5.4 Baryons and the Eightfold Way.

about 150 MeV. For this reason Gell-Mann supposed that the particle with strangeness
-3 would be another 150 MeV heavier than the strangeness -2 Ξ^*: thus he predicted
approximately 1680 MeV for it.

In 1963 at Brookhaven Laboratory in New York, and independently at CERN,
Geneva, the predicted particle was found. Its strangeness was -3, its charge was
negative, and its mass was 1679 MeV.

The Omega Minus was the final link in the patterns and established their relevance.
This was as significant as Mendeleev's table of the atomic elements had been. With
this successful prediction of new elementary particles (Figure 5.5), Gell-Mann had
paralleled Mendeleev's prediction of the atomic elements Gallium, Germanium, and
Scandium.

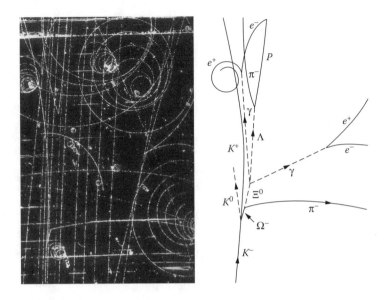

FIGURE 5.5 The discovery of Ω. (Courtesy of American Physical Society and Dr. D. Radojicic.)

With the validity of the Eightfold Way established, the crucial question that it posed was — why? What is the cause of this pattern underlying the abundance of supposedly elementary particles? The importance of resolving this grew as more and more particles were found (in fact, over a hundred species of particles were discovered in the quarter-century following the discovery of the pion back in 1947). It was with the discovery of this multitude that the riddle was solved.

6 Quarks

QUARKS AND THE EIGHTFOLD WAY PATTERNS

With hindsight, it is possible to pinpoint Mendeleev's periodic table of the atomic elements in 1869 as the first hint of a more fundamental layer of matter common to all atoms, and responsible for giving them their properties. Half a century later, the discovery that atoms consisted of electrons encircling a nucleus confirmed this. By the middle of the 20th century, the structure of the nucleus was in its turn being revealed. The observation of a recurring pattern among the 30 or so hadrons known in the early 1960s was an analogous pointer to the possibility of a more fundamental variety of matter — *quarks* — out of which these hadrons, including the neutron, and proton, and ultimately the nucleus, are formed.

As early as 1964, Murray Gell-Mann and George Zweig independently noticed that the Eightfold Way patterns would arise naturally if all the known hadrons were built from just three varieties of quark. Two of these, known as the 'up' and 'down' quarks (u and d for short), are sufficient to build the baryons that have zero strangeness.[1] Strange hadrons contain the third variety, the 'strange' quark (s for short). The more strange quarks there are present in a cluster, the more strangeness the cluster has. (See Figures 6.1 and 6.2.)

Quarks are unusual in that they have electrical charges that are fractions of a proton's charge: the up (u) has charge $+2/3$ and the down (d) $-1/3$. As no one had seen direct evidence for an isolated body with fractional charge, Gell-Mann and Zweig's idea initially received a mixed reaction. However there was no denying that the idea worked and although no one has ever liberated a quark from a proton, quarks with these charges have been detected *inside* the proton, confirming their reality. If you form a group of three quarks, each one being any of the up, down, or strange varieties, then the Eightfold Way pattern of baryons emerges.

Before illustrating how this happens, let me answer some questions that may have entered your mind: Why clusters of *three*? Why not two, or five? Why not individual quarks?

These very questions were asked insistently by many physicists in the latter half of the 1960s. All that one could then reply was: 'Because it seems to work that way,' and hope that a more complete answer would eventually be forthcoming. Such an answer did subsequently emerge, rationalising all of the assumptions, and will appear in Chapter 7. I could at this point have chosen to jump forward here to the discoveries of the 1970s, which confirmed the reality of quarks, and then backtracked to 1964 — presenting Gell-Mann and Zweig's ideas as if they had the benefit of foresight. If you wish to follow that route, proceed to p. 85 first, but that is not how things

---

[1] There are of course many mesons with zero strangeness also made from up or down quarks and antiquarks. However, there are also mesons with zero strangeness made of a strange quark and a strange antiquark where the strangeness of the quark and the antiquark mutually cancel; see Table 6.2.

If quarks occur in two varieties: up or down, then all of the material in the world about us can be built.

The proton (charge +1)

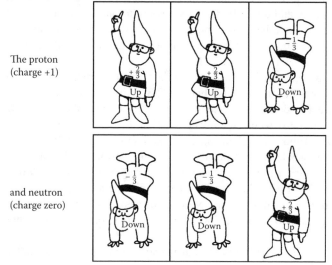

and neutron (charge zero)

build the nuclei of all atoms. The pions that attract them to one another are formed from up and down quarks and antiquarks.

FIGURE 6.1 A pair of quarks. (*Continued.*)

developed historically. I am using the benefit of hindsight to edit out the red herrings and false trails that always plague research at the frontiers, and so I might inadvertently give the impression that progress is inexorably forwards and free from uncertainty. In practice it is not like that. For proponents of the quark theory, the latter half of the

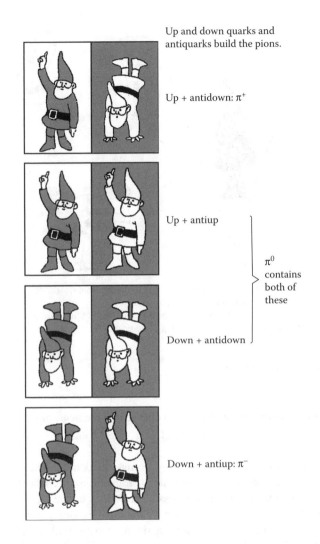

Up and down quarks and
antiquarks build the pions.

Up + antidown: π^+

Up + antiup

π^0
contains
both of
these

Down + antidown

Down + antiup: π^-

FIGURE 6.1 (Continued).

1960s was an eerie interregnum when they were able to reach correct conclusions
through reasoning that had little or no good foundation, based on weird particles that
no one had even seen.

Putting aside this and other justified questions temporarily, suppose that I cluster
three quarks together in any combination of up, down, or strange. Adding together
the electric charges of the quarks gives the total charge of the cluster. Thus, two ups
and one down will have the same charge as a proton:

$$u^{(+2/3)} + u^{(+2/3)} + d^{(-1/3)} = p^{(+1)}$$

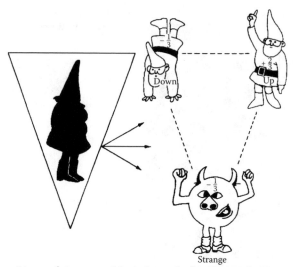

The existence of strange particles is due to the fact that quarks can
occur in three varieties: up, down, or strange.
Quark plus antiqaurk now yield nine possibilities, and there are strange
partners of proton and neutron (e.g. Σ sigma and Λ lambda particles).

FIGURE 6.2 A triangle of quarks.

while two down and one up have net zero charge like the neutron:

$$d^{(-1/3)} + d^{(-1/3)} + u^{(+2/3)} = n^{(0)}$$

Strangeness is a property possessed by strange quarks: the more strange quarks
that are present in a cluster, the more amount of strangeness the cluster will have. (The
neutrons and proton have zero strangeness because they contain no strange quarks.)
Furthermore, if the down and up quarks have identical masses and the strange quark
is 150 MeV heavier, then one can understand why clusters with a lot of strangeness
are heavier than their siblings with less strangeness. The Ω^- with strangeness -3
consists of three strange quarks and is 150 MeV heavier than the Ξ^* (strangeness -2)
which is in turn 150 MeV heavier than Σ^* (strangeness -1), and this is yet another
150 MeV heavier than the zero strangeness Δ particles.

(The historical accident that strangeness entered the language of physics before
the quark model and strange quarks were thought of, has led to the topsy-turvey
accounting that a strange quark has one unit of *negative* strangeness. Hence the Ω^-
containing three strange quarks has strangeness -3. It is too late to change this,
regrettably.)

To see how this all works, form all possible clusters of three quarks and tabulate the
sum of their electrical charges and strangeness using the individual quark properties
listed in Table 6.1. We find the following:

Clusters with baryon number = + 1	Strangeness = − number of strange quarks	Charge = sum of quark charges	Examples		
uuu		2	Δ		
uud	0	1	Δ	p	
udd		0	Δ^0	n	
ddd		− 1	Δ^-		
uus		1	Σ^{*}	Σ	
uds	− 1	0	Σ^{0*}	Σ^0	Λ^0
dds		− 1	Σ^{-*}	Σ^-	
uss	− 2	0	Ξ^{0*}	Ξ^0	
dss		− 1	Ξ^{-*}	Ξ^-	
sss	− 3	− 1	Ω^-		

Table 6.1 Quarks

Flavour	Electrical charge	Strangeness
u	$\frac{2}{3}$	0
d	$-\frac{1}{3}$	0
s	$-\frac{1}{3}$	−1
$\bar{u}$	$-\frac{2}{3}$	0
$\bar{d}$	$+\frac{2}{3}$	0
$\bar{s}$	$+\frac{1}{3}$	+1

Charge and strangeness of the up, down, and strange flavours of quarks. The $\bar{u}\bar{d}\bar{s}$ antiquarks have opposite values for charge and strangeness compared to the u,d,s quarks.

The column of ten corresponds exactly with the decuplet of particles that contains the Ω^- (Figure 6.3b). If we take clusters where at least one quark differs from the other pair, then we find the eight members of the octet that contains the proton (that it is eight and not seven is a subtlety arising from the fact that in the (uds) cluster, all three are distinct).

The up, down, and strange properties are collectively referred to as the 'flavours' of the quarks. With these three flavours we have constructed the Eightfold Way patterns for the baryons. For each and every flavour of quark there is a corresponding antiquark having the same mass and spin as its quark counterpart but possessing opposite sign of strangeness and charge. Thus, while the strange quark, s, has charge $-1/3$ and strangeness -1, the strange *anti*quark, denoted $\bar{s}$, has charge $+1/3$ and strangeness $+1$. By clustering together three antiquarks we obtain the antibaryon counterparts to the octet and decuplet baryons.

(a) The families of eight and ten baryons, and
(b) The quark systems that generate them.

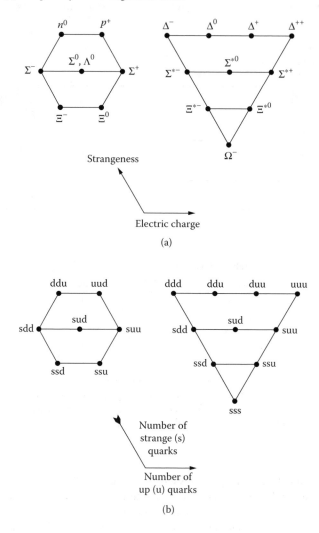

FIGURE 6.3 Baryons.

With both quarks and antiquarks available, there is another way that we can build clusters with integer charges: form a system containing a single quark and an antiquark. From the three quark flavours and three antiquark flavours in Table 6.1, we can form nine possible combinations: $u\bar{d}$, $u\bar{u}$, $d\bar{d}$, $s\bar{s}$, $d\bar{u}$, $u\bar{s}$, $d\bar{s}$, $s\bar{u}$, $s\bar{d}$. The electrical charge and strangeness of each of these is obtained by adding together the quark and antiquark contribution as in the table. Astonishingly, these are precisely the combinations of charge and strangeness that mesons such as π, K, η are found to have (see Table 6.2). This is a profound result. For example, there are no mesons with

Table 6.2 Meson Nonets

Clusters with zero baryon number	Strangeness = number of strange antiquarks – number of strange quarks	Charge = sum of quark charges	Examples
$u\bar{s}$	+1	+1	K^+
$u\bar{d}$		0	K^0
$u\bar{d}$		+1	π^+
$\left.\begin{array}{l} u\bar{u} \\ d\bar{d} \\ s\bar{s} \end{array}\right\}$	0	0	π^0, η^0, η'^0
$d\bar{u}$		−1	π^-
$s\bar{u}$	−1	−1	K^-
$s\bar{d}$		0	$\bar{K}^0$

strangeness −2, whereas such baryons do occur, and strangeness −1 states can have charge +1 for baryons but not for mesons. This is precisely what happens in nature and is easily explained by the quark model, as in Figure 6.4.

If quarks are real particles then they will have other properties that will be manifested in the hadrons formed from their clusters. For example, quarks spin at the same rate as do electrons, namely, a magnitude of 1/2 (in units of Planck's constant $\hbar = h/2\pi$). Since an odd number of halves gives a half-integer, then a three quark system has half-integer spin — precisely as observed for baryons. Conversely, an even number of halves gives an integer — quark plus antiquark have integer spins, as do mesons, thus explaining the observation in Table 5.4.

The rules for adding up spins (Figure 6.5) had been known since the advent of quantum mechanics 40 years before these discoveries. It had been applied first to electrons in atoms, then later to nucleons in nuclei, and so can confidently be applied to quarks in hadrons, as follows.

Two spin 1/2 objects, such as quark and antiquark, combine to a total of 0 or 1. Indeed, nine mesons with spin 0 are known — the familiar set containing π and K.

The first hints of the Eightfold Way pattern among the hadrons (strongly interacting particles) came from the family of eight baryons containing the proton, and a similar hexagonal pattern with seven mesons (π, K) known in the 1950s. To complete the correspondence between the two, an eighth meson (η) was predicted. Its subsequent discovery gave much support to the scheme. Later a ninth meson was discovered (η') which breaks the direct correspondence. Today we recognise that baryons like these occur in families of eight or ten but mesons occur in nonets (nine). This emerges naturally in the quark model where baryons are clusters of three quarks whereas mesons are clusters of quark and antiquark. This is one of the model's many successes.

The triangle occurs if charge is plotted on a skewed axis and strangeness on the vertical axis.

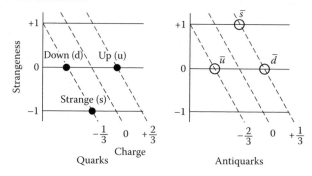

Quarks   Antiquarks

The nine quark plus antiquark possibilities arise if an antiquark triangle is drawn centred on each vertex of a quark triangle.

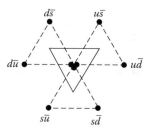

This yields the hexagon pattern for mesons. Three states are at the centre: $u\bar{u}$, $d\bar{d}$, $s\bar{s}$.

FIGURE 6.4 Quark triangles and the hadronic particle patterns.

Nine spin 1 mesons also exist, consisting of ρ^-, ρ^0, ρ^+ with masses of 770 MeV; and ω^0 of about the same mass; K^{*-}, K^{*+}, K^{*0}, $\bar{K}^{*0}$ masses 890 MeV; and the ϕ^0, mass 1020 MeV completes the family nicely. Indeed, it was the discovery of these spin 1 mesons, in particular the ϕ in 1963, that played an essential role in the development of the quark hypothesis.

Three spin 1/2 objects, such as three quarks, combine to a total spin of 1/2 or 3/2. The eight members of the family containing the proton each have spin 1/2; the ten containing the Δ, Σ^*, Ξ^*, and Ω^- each have spin 3/2. The successful explanation of spin = 0 or 1 for mesons built from quark plus antiquark has been matched by the spin = 1/2 or 3/2 for baryons built of three quarks.

AN 'ATOMIC' MODEL OF HADRONS

Atoms consist of electrons in motion about a central nucleus. The nucleus itself has an internal structure consisting of neutrons and protons. The spinning and orbiting motions of the electrons in atoms or the nucleons in nuclei give rise to assorted excited

When adding together two or more angular momenta, we must take account of their vector character: the direction of spin is important in addition to its magnitude.

In subatomic systems, the angular momentum is constrained to be an integer multiple of Planck's quantum $\hbar$ $h/2\pi$

$$L \quad n\hbar \qquad (n \quad 0, 1, 2\ldots; \text{known as S, P, D}\ldots\text{states})$$

The sum of two angular momenta must itself be an integer multiple of $\hbar$. Thus, depending upon the relative orientations of L_1 and L_2, the sum can have any value from $(n_2 + n_1)$; $(n_2 + n_1 - 1)\ldots$ to $|n_2 - n_1|$. (Each of these is understood to be multiplied by $\hbar$ but conventionally the $\hbar$ is often omitted and so we write L 3 not $3\hbar$.)

Electrons have an intrinsic angular momentum or 'spin' of magnitude $\frac{1}{2}\hbar$. Adding to L $n_1\hbar$ gives $(n_1 + \frac{1}{2})\hbar$ or $(n_1 - \frac{1}{2})\hbar$. The difference of these is an integer multiple of $\hbar$. In general, adding $n_1\hbar$ to $\frac{m\hbar}{2}$

(m odd or even) gives:

$$\left(n_1 + \frac{m}{2}\right); \left(n_1 + \frac{m}{2} - 1\right)\cdots \left|\left(n_1 - \frac{m}{2}\right)\right|$$

as the set of possibilities.

Some specific examples may help make the point. Two spin $\frac{1}{2}$ will add to either 1 or 0. Three spin $\frac{1}{2}$ will yield $\frac{3}{2}$ of $\frac{1}{2}$. Two $L = 1$ will add to yield total 2, 1, or 0.

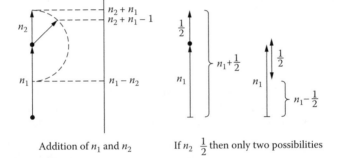

Addition of n_1 and n_2 If n_2 $\frac{1}{2}$ then only two possibilities

FIGURE 6.5 Angular momentum.

states of atoms and nuclei. Thus if hadrons are clusters of quarks, we should expect by analogy that excited hadronic states will occur as a result of the various spinning and orbiting motions of their constituent quarks.

When the quarks have no orbital motion about one another, the total spin of the cluster comes entirely from the spins of the individual quarks within. We have already seen how this gives spin 0 or 1 for mesons and spin 1/2 or 3/2 for baryons. What happens if we now admit orbital motion for those quarks in addition to their intrinsic spins?

The total spin of the hadronic cluster will result from the quarks' spins and also their mutual orbital angular momenta. The more orbital motion the quarks have, the larger the total spin of the hadron will be. Quarks in a state of rapid orbital motion carry more energy than when orbiting slowly, thus the energy or mass of a hadron with large spin will tend to be higher than that of hadrons with small spins.

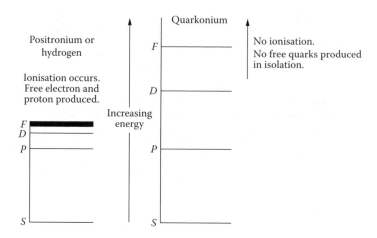

The essential difference between the electromagnetic force and the force that binds quarks is that electric charges can be liberated from their binding in atoms (e.g., by heat) whereas quarks seem to be permanently confined inside their atomic clusters (hadrons). Apart from this, the pattern of levels, correlation of increasing mass and spin, splitting of masses for parallel from antiparallel spin are all very similar.

FIGURE 6.6 Pattern of energy levels in atoms and quark clusters.

This is indeed the case in nature. Hundreds of hadrons exist, and the higher their spin, the larger their masses are seen to be. There are so many that no one can memorise all of their properties; these are listed and revised biennially in a publication of increasing bulk. Fortunately we do not have to know them all. Fermi is reputed to have said that if it were necessary to know the names of all the hadrons then he might as well have been a botanist. We can easily summarise them though. These hundreds of particles form the ubiquitous hexagonal families of the Eightfold Way, the spin 0 and 1 mesons being accompanied by spin 2, 3, and 4 mesons. Baryon patterns have been observed with spin 1/2, 3/2, 5/2 and so on up to 15/2 (so far!). The Eightfold Way patterns bear testimony to the quark constituents, the increasing spins, and masses exhibiting the dynamic motion of those quarks within the hadrons.

The picture we have today is that hadrons are clusters of quarks much as atoms are clusters of electrons and nuclei. There are tantalising similarities here, but also some profound differences.

When energy is supplied to a hydrogen atom, the electron is raised into states of higher angular momentum and energy. If enough energy is supplied, the electron will be ejected from the atom and 'ionisation' occurs. The energy to excite electrons from the ground state ("S-state") to higher energy states (such as the P-state) is of the order of a few electron-volts.

Compare this with quark clusters. To excite a quark requires hundreds of MeV. The mesons π and ρ, where the quark and antiquark are in the lowest energy state (the S-state in the language of atomic physics) have masses of 140 and 770 MeV, respectively; the masses of their P-state counterparts range from 1000 to 1400 MeV.

This is in part due to the fact that mesons are much smaller than atoms and thus typical energies are correspondingly greater (the uncertainty principle underwrites microscopic phenomena — small distances correspond to high momentum or energy, and vice versa). It is also due to the nature of the forces amongst quarks, which are much stronger than electromagnetic forces and so provide more resistance to excitation.

The other noticeable feature is that although the patterns of increasing energy with increasing spin are essentially the same in quark clusters and atoms, the relative energy gaps between analogous configurations in the two cases are quite different (Figure 6.6). In hydrogen, the amount of energy required to excite the electron from the S to P configuration is already nearly enough to eject it from the atom. In quark clusters, things are not like this. The separation of S to P configurations is roughly the same as P to D and so on. As energy is supplied to a quark cluster, the quarks are excited to a higher energy configuration but are not ejected from the cluster — there is no analogue of ionisation. Quarks are said to be "confined" within hadrons.

QUARKS IN THE PROTON

By the late 1960s more than a hundred species of hadrons had been found. Each and every one of these had properties suggesting that they were built from quarks as in the above scheme. The accumulating evidence convinced most physicists that the quark hypothesis was probably correct even though individual quarks had not been found; quarks had only manifested themselves in their clusters — the subnuclear particles: hadrons.

Many searches for isolated quarks were made. Their most dramatic property is that they are required to have electrical charges of 2/3 and $-1/3$, whereas all particles ever seen and confirmed have integer or zero charges. This possession of a fractional electric charge should make an isolated quark very obvious. It was suggested by some that quarks might be so massive that no accelerator on Earth was powerful enough to produce them. However, one would have expected them to be produced in collisions between cosmic rays and the atmosphere and even if this happened only rarely, their fractional charges would be so distinctive that it would be difficult not to notice them.

With the failure of the early quark searches, suspicion grew that quarks were not really physical objects but were somehow mathematical artifacts that conveniently produced the patterns of the particles while having no real significance in themselves.

The turning point came as a result of a series of experiments at SLAC, the Stanford Linear Accelerator Center in California, from 1968, and at CERN, Geneva from 1970 which saw direct evidence for quarks physically trapped inside the proton and neutron. The significance of these experiments parallels Rutherford's 1911 discovery of the atomic nucleus and they were in essence simply more powerful versions of that experiment. Indeed there are many similarities in the revelation of atomic substructure at the start of the 20th century and the uncovering of the quark layer of matter towards its end.

Early in the 1950s, collisions of protons with other protons had shown that they had a diameter of about 10^{-15} metres, small compared to the size of a nucleus but

(a)

(b)

FIGURE 6.7 Stanford Linear Accelerator (SLAC) of electrons in California. (a) The back of the tube is located at point A; electrons accelerate along the tube (2 miles long), and the 'screen' consists of detectors in the building at point B. (b) Electron detectors at point B. (Courtesy of Stanford University.)

more than a thousand times that of an electron. With this discovery that the proton was 'large,' suspicion arose that it might have an internal structure.

The original guess was that the proton's size was due to a cloud of pions perpetually surrounding it. In this picture, the proton was pregnant with the carriers of the nuclear force by which neighbouring protons or neutrons could be ensnared and nuclei formed. Although appealing, this failed to give a quantitative description of nucleon properties such as magnetic moments, nor did it fit easily with the Eightfold Way patterns of the hadrons.

After the quark hypothesis first appeared in 1964, the idea gained ground that quarks in motion gave the proton its size, perhaps in analogy to the way that electrons and nuclei gave size to atoms.

In the 1960s, a 2-mile long machine was built at Stanford in California capable of accelerating *electrons* until they had energies in excess of 20 GeV (see Figure 6.7). At these high energies, electrons can resolve structures less than 1 fermi (10^{-15} m) in size and are therefore a perfect tool for probing inside protons and investigating their structure.

The electron's negative charge causes it to be attracted or repelled, respectively, by the up and down quarks which have electrical charges $+2/3$ and $-1/3$. The quarks' spinning motion causes them to act as magnets which exert calculable magnetic forces on the passing electrons. Taking all of these things into account, it is possible to predict what should happen when an electron beam hits a proton at high energy. You can calculate the chance that it is scattered through some angle, how much energy it loses while the proton recoils, and so on.

By firing beams of high-energy electrons at targets of protons (for example, liquid hydrogen) and observing how they scatter, you can determine where the charge of the proton is concentrated. If it was evenly distributed throughout the whole volume then the proton would be seen as a diffuse cloud of electricity and the electron beam would pass through with little or no deflection. However, if the charge is localised on three quarks then the electron will occasionally pass close to a concentration of charge and be violently deflected from its path, analogous to Geiger, Marsden, and Rutherford's experiments (using α particles instead of electrons) that revealed the nuclear atom in 1911.

Violent collisions *were* seen at SLAC, and the distribution of the scattered electrons showed that the proton is indeed built from entities with spin 1/2 such as quarks. Comparison of these results with similar experiments at CERN (where neutrinos were used as probes in place of electrons) showed that these constituents have electrical charges which are $+2/3$ and $-1/3$ fractions of the proton's charge. These are identical to the properties that had been deduced for quarks from the Eightfold Way patterns, and confirm the quarks' presence in the proton.

Some additional discoveries about the inside of the proton were made. First of all, the experiments showed that electrically neutral particles ('gluons') exist there in addition to quarks. Just as photons are the carriers of the electromagnetic force between electrically charged particles, so it was suspected that these gluons might be carriers of the force that attracts quarks together forming the proton. Today we know this to be correct (see Chapter 7).

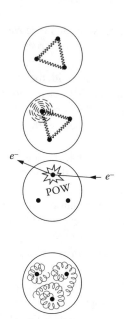

The proton can be viewed as three quarks very tightly bound by super-strong force.

Exciting one or more quarks yields baryon resonance states like Δ.

SLAC, California (Figure 6.7). High energy electrons scatter from quarks in the proton, giving the first direct evidence for quarks. Paradox: quarks appear to be free!

Neutrinos scatter from protons at CERN. Comparison with electron scattering reveals that in addition to the quarks there is also electrically neutral material inside the proton, dubbed 'gluons.' Suspicion arises that this may be the glue that holds the quarks to one another in the proton.

1970s Proton: quarks held by gluons

Paradox: Gluons bind quarks very weakly; but try to get a quark out and the glue comes on strong.

Theorists search for a theory of quark force that will have these properties (see Chapter 7).

FIGURE 6.8 The quark force paradox (late 1960s).

The discovery of gluons was most welcome as it gave a strong hint that we were indeed revealing the inner workings of the proton, not just what it is made from but how it is held together. However, the way that the electron beams scattered from the quarks within the proton gave some surprises. First, the quarks turned out to be very light, less than one third of a proton mass. Second, they appeared to be almost free inside the proton as if they are hardly glued together at all!

If this is really what quarks are like then you would expect them to be easily ejected from the proton. Indeed, as soon as these phenomena were seen in 1968, plans were made to see what happened to the proton after it had been struck so violently. For a year or so there were hopes that individual quarks might emerge, but these hopes were short lived — pions and other familiar particles were produced but no free quarks or gluons appeared.

Although this was a disappointment for the experimentalists, it created an exciting paradox for the theorists (Figure 6.8). As one physicist succinctly put it, 'The proton is like an ideal prison: the quarks have perfect freedom inside but are unable to escape.' It is analogous to quarks being held to one another by an elastic band that is slack. The quarks are apparently free, but after being struck they recoil and the elastic becomes tighter, preventing them from escaping. The elastic may become so stretched that it snaps. The two new ends created have quarks on them, and so mesons are formed but not free quarks.

This paradox was seminal in the subsequent development of theoretical physics. The theory of free quarks that are nonetheless permanently bound is known as 'Quantum Chromodynamics' or QCD and its development and present status are the next topics in this story.

7 Quantum Chromodynamics: A Theory for Quarks

COLOUR

The discovery that protons and neutrons are built from quarks and the earlier discoveries of atomic and nuclear structure share some interesting parallels. There are four common ingredients whose psychological impacts have been rather different in each case, primarily because the sequence in which they came to the fore differed.

1. "Fundamental" objects, supposedly independent of one another and structureless, nevertheless exhibit common properties. These features are distributed or repeated among the objects in a regular fashion such that they exhibit a definite pattern. Examples of such patterns are the Periodic Table of atomic elements and the "Eightfold Way" of hadrons. These regularities hint that the supposedly elementary featureless objects are built from more basic common constituents.

2. When beams of particles (such as electrons, neutrinos, or alpha particles) are fired at the objects of interest, the beams are scattered. The violence of this scattering shows that the supposedly featureless objects in the target contain a complex inner structure, and are built from more fundamental constituents.

3. Attractive forces are identified which bind the constituents to one another, forming the more complex structures.

4. Pauli's exclusion principle plays a crucial role, governing the ways that the constituents can link together. This limits the number and form of the resulting complex structures such that the allowed ones exhibit common properties and patterns of regularity. Pauli's exclusion principle is most familiar in atoms, where it forbids more than one electron from occupying the same energy state and is thereby responsible for generating the regular pattern of the atomic elements as discerned by Mendeleev. Pauli's principle applies also to quarks and controls the ways that they can combine. This results in the patterns of the Eightfold Way.

In the case of the atomic elements, the electron and nuclear structures of their atoms were identified long after the electromagnetic force had been understood. So scientists were already familiar with the forces at work inside atoms, though it was not until the formulation of quantum mechanics in the 1920s that the electronic structures of atoms were fully understood. (See Table 7.1.)

For quarks and elementary particles, the sequence of events was very different. Ingredients 1 and 2 had been recognised in the 1960s with the discovery of the Eightfold Way patterns of hadrons, and the observation that beams of electrons or

Table 7.1 A Summary of our Knowledge in the Late 1960s

	Atom	Nucleus	Hadron
Pattern	Mendeleev table 1869	Isotopes and magic numbers known early 20th century	Eightfold way 1962
Constituents identified	α-Particle scattering	α-Particle scattering	Electron and neutrino scattering
	↠ Nucleus 1911 ionisation ↠ Electrons	↠ Proton 1919 . . . neutron 1932	↠ Quarks 1968–70
Clustering force	Electromagnetic force (already known)	Strong force (inferred 1935)	? (Unknown)
Force Carrier and theory	Photon Quantum electro-dynamics 1948	Pion Yukawa model	 ? (Unknown)
Pauli Principle examples	Electrons occupy energy 'shells' ↠ Chemical regularity ↠ Mendeleev table	At most two protons and two neutrons in lowest energy state ↠ α-Particle stable ↠ Isotopes	Forbids three identical strange quarks to simultaneously occupy lowest energy state ↠ Ω⁻ cannot exist ? (Paradox)

neutrinos scatter violently from the quarks within those hadrons. Nothing was known about the forces that act on quarks beyond the fact that quarks have a strong tendency to cluster in threes (for example, forming baryons) or for a single quark to bind with an antiquark (meson). These were empirical facts that any theory of quark forces would have to explain, but in the absence of other clues would not be sufficient to lead to that theory. The first real clue, though none saw its full significance at once, was a paradox concerning the Pauli exclusion principle.

Spin 1/2 particles (such as electrons, protons, neutrons, and quarks) obey this principle which forbids more than one of a given kind being in the same state of energy and direction of spin. A familiar example is in the formation of atoms, where Pauli's principle forces electrons into particular orbital configurations and, as a result, a periodically repeating pattern of chemical properties occurs, as originally noted by Mendeleev. In nuclear physics the principle allows at most two protons and two neutrons in the lowest energy state. This is the source of the stability of the α particle, helium-4, and leads to elements such as oxygen and iron being as common as dirt while gold is a relative rarity.

Quarks have spin 1/2 and so the principle should apply to them, too. A spin 1/2 particle can spin in either of two directions — clockwise or anticlockwise — allowing at most two quarks to have the same energy if they have identical flavours. It is natural to expect that the lightest clusters are formed when each quark is in its lowest energy state, thus the Ω^-, which consists of three identical strange quarks, is seemingly forbidden to exist contrary to clear evidence that it does!

One strange quark spinning clockwise.

Second strange quark. It must spin the opposite way so that it is distinguishable from the first.

Third quark has only two possible ways to spin—clockwise or anticlockwise. But both are forbidden as there are already quarks present in these states.

FIGURE 7.1   The Ω^- problem. (*Continued.*)

Oscar (Wally) Greenberg recognised this problem with the Pauli principle in 1964 soon after the idea of quarks had been proposed. To resolve it, he suggested that quarks possess a new property called "colour," which we now recognise is in many ways similar to electric charge except that it occurs in three varieties (Figure 7.1). To distinguish among these, particle physicists have whimsically referred to them as the red, yellow, or blue variety, known collectively as 'colour' charges. Instead of simply positive or negative charge, as is the case for electric charge, there are positive or negative 'red,' 'yellow,' or 'blue' colours. Quarks carry positive colour charges and antiquarks have the corresponding negative colour charges. Thus a strange quark can occur in any of these forms and, to distinguish them, we append the appropriate subscripts s_R, s_B, s_Y. Similarly, the up and down quarks can be u_R, u_B, u_Y and d_R, d_B, d_Y. For antiquarks we have $\bar{s}_R, \bar{s}_B, \bar{s}_Y$ and so on (Figure 7.2).

Pauli's principle only forbids *identical* quarks to occupy the same spin and energy state. Thus if one of the strange quarks in the Ω^- carries the red variety of charge, while one has yellow and the other one blue, then they are no longer identical and so the Ω^- can exist as empirically observed.

Given that before 1968 the quark idea was still controversial, it is quite remarkable that Greenberg did not take the Pauli paradox as evidence that the quark model was

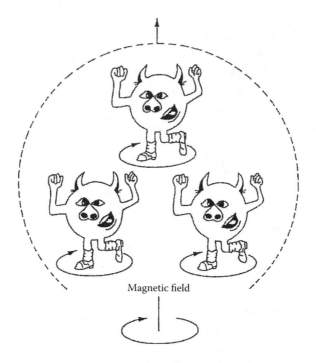

Ω^- is seen to exist and spin in a magnetic field with each quark spinning clockwise.
How can this be?
Quarks must possess some further property that enables them to be distinguishable.

If strange quark can exist in any of three colours, we can distinguish each one in the Ω^-.

FIGURE 7.1 (Continued).

wrong, but instead proposed the property of colour to overcome the problem. The chief merit of the quark model in 1964 was that it provided a simplification; his idea of introducing a new property that in effect multiplied the number of quarks by three created little enthusiasm at first. However, attitudes began to change around 1970 following the experiments that scattered electron beams from protons and neutrons, and which showed that they are indeed made of quarks. Although these experiments showed that quarks were present, they were incapable of showing whether or not quarks

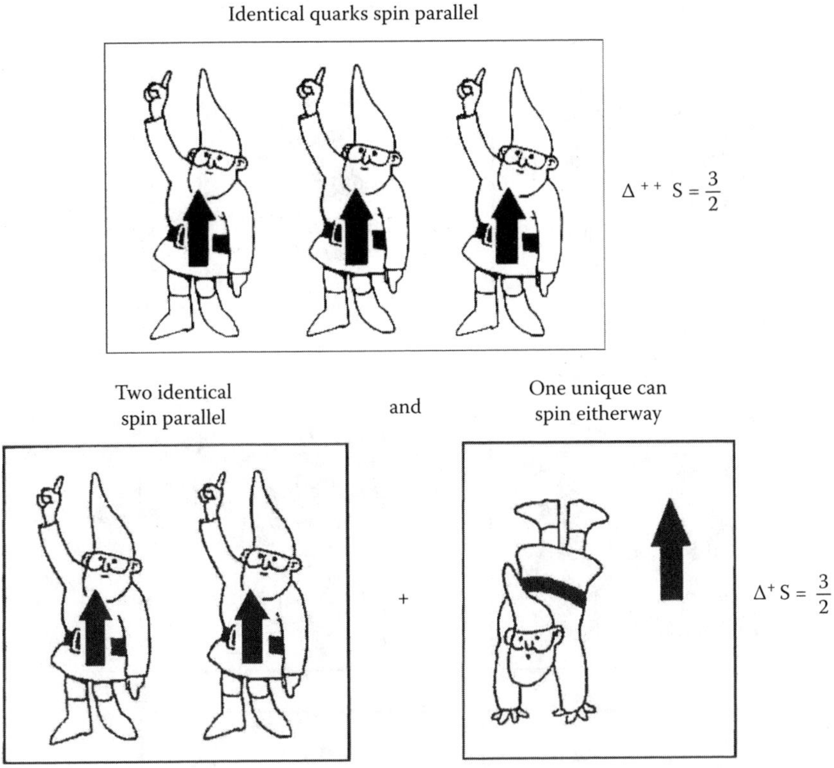

Identical quarks spin parallel

$\Delta^{++}\ S = \frac{3}{2}$

Two identical   and   One unique can
spin parallel   spin eitherway

$+$

$\Delta^{+}\ S = \frac{3}{2}$

FIGURE 7.2 Colour, the Pauli principle, and baryon spins. The Pauli principle forbids two identical spin $\frac{1}{2}$ particles to occupy the same state of energy and spin. Thus the two electrons in the ground state of the helium atom must spin in opposite directions–antiparallel. Similarly, the two protons in an α particle must spin antiparallel as must the two neutrons.

The same would be true for quarks if they did not possess colour which distinguishes the otherwise identical strange quarks in the Ω^- or the two up quarks in the proton, for example. The effect of colour combined with the Pauli exclusion principle is that any two quarks having the same flavour (two up, two down, two strange) in the lowest energy state must spin *parallel* — precisely the opposite of what happens in atoms and nuclei.

An extreme example is when all three quarks have the same flavour, as in Δ^{++} (uuu), Δ^- (ddd), or Ω^- (sss). Here all three quarks must spin parallel, hence the total spin is $\frac{3}{2}$ (Figure 7.1).

If two flavours are identical and the third differs (e.g., Δ^+ (uud) or p (uud)) then the identical pair must spin parallel but the third is not constrained, it can spin parallel (hence total spin $\frac{3}{2}$ as in the Δ) or antiparallel (hence total spin $\frac{1}{2}$ as in the proton).

Thus we see that the decuplet containing Δ^{++}, Δ^-, Ω^- naturally has spin $\frac{3}{2}$ as observed. Removing the cases where all three quarks are identical leads to the octet where the total spin is $\frac{1}{2}$ precisely as in nature. When all three quarks have different flavours (uds), then any pair can be spinning parallel or antiparallel, hence the extra state at the uds site in Figure 6.3. The $\sum^{*0}$ has the ud and s all parallel; the $\sum^0$ has ud parallel and opposite to s while the Λ^0 has ud antiparallel. (*Continued.*)

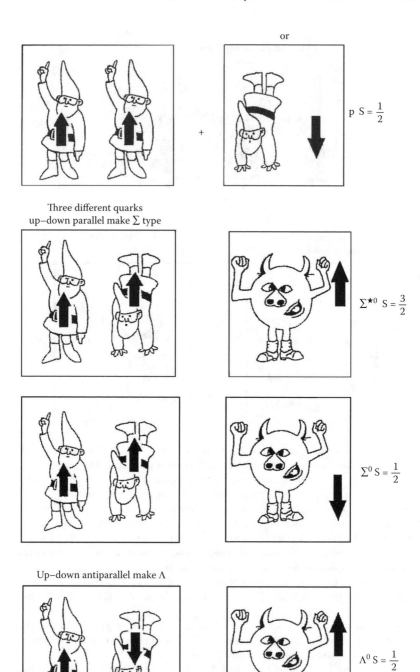

FIGURE 7.2 (Continued).

had colour. The crucial evidence for *coloured* quarks came instead from experiments where electrons collide with their antimatter, positrons, and mutually annihilate.

When an electron and positron annihilate, the energy of their motion is converted into new varieties of matter and antimatter such as a muon and antimuon or quarks and antiquarks. The quarks and antiquarks cluster together, forming the familiar hadrons such as protons and pions, and it is these that are detected. The probability that hadrons are produced relative to that for muons and antimuons to emerge is given by the sum of the charges squared of all varieties of quarks that are confined inside those hadrons. The relative abundances of the flavours then known (u, d, s) was predicted to be:

$$\frac{\text{Production rate of hadrons}}{\text{Production rate of muon and antimuon}}$$

$$= \left[(2/3)^2_{(up)} + (-1/3)^2_{(down)} + (-1/3)^2_{(strange)} \right] \times 3_{(\text{if three colours})}$$

hence the result is 2/3 if quarks do not have colour (like leptons) but 2 if they occur in three colours. This experiment was performed in 1970 at Frascati near Rome. The ratio was seen to be much larger than 2/3 and consistent with 2 within experimental uncertainties (though to be historically accurate I should stress that the experimental uncertainties were rather large. It was not until these experiments were reproduced with greater precision at Stanford, California after 1972 that evidence for colour started to become convincing). At last, here was direct evidence supporting the notion that there are three distinct colours of up quark, three down and three strange. Today we know also of charm, bottom, and top flavours (Chapter 9), each of which also occurs in three colours.

COLOUR AND QUARK FORCES

Quarks, and hadrons containing quarks, all experience the strong nuclear force, whereas electrons and neutrinos do not. As soon as quarks were discovered to have colour, a property that electrons and neutrinos do not have, the idea began to take hold that colour might be the source of the forces acting between quarks. If this was correct then it would explain naturally why electrons and neutrinos are blind to the strong nuclear forces.

How shall we build a theory of colour? The inspired guess made in 1972 was that colour interactions are analogous to the interactions among electric charges. In electrostatics, like charges repel and opposite charges attract and the analogy for colour is immediate: like colours repel and opposite colours attract. Thus two red quarks repel one another but a red quark and an 'anti-red' antiquark will mutually attract. Similarly, blue attracts anti-blue or yellow attracts anti-yellow. This is very encouraging because it explains the existence of mesons naturally — just as positive and negative electrical charges attract to form net uncharged atoms, so have positive and negative colours, carried by quark and antiquark, attracted to form net uncoloured hadrons.

Colour and opposite colour attract. Thus is formed a **meson** $q\bar{q}$

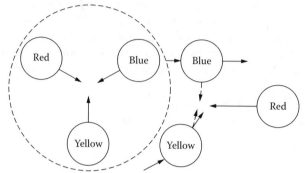

Three different colours attract. Nearby like colours are repelled.
Colourless clusters of three different coloured quarks form. Hence
baryons $q_R q_B q_Y$.

FIGURE 7.3   Colour attractions form mesons and baryons.

In electrostatics, two positive charges are always 'like charges' and repel. For
colour, two red quarks are always coloured alike and repel, but what about a red quark
and a blue quark? These are alike in that they are both quarks ('positive colours') but
unlike in that the colours differ.

It turns out that these different colours can attract one another but less intensely
than do the opposite colours of quark and antiquark. Thus, a red quark and a blue
quark can mutually attract, and the attraction is maximised if in turn they cluster with
a yellow quark (Figure 7.3). Red and yellow, red and blue, blue and yellow all attract
one another and thus do the three quark clusters known as baryons occur. Notice
that the baryon formed in this way necessarily contains three quarks, *each one with
a different colour*. Thus, we have been led naturally to the picture that Greenberg
invented ad hoc in 1964 as an explanation of the Pauli exclusion paradox.

The mathematics that has been developed to describe colour interactions shows
that the above clusterings — quark and antiquark of opposite colours or three quarks
each of different colour — are the simplest ways that net uncoloured hadrons can be
formed. Nature seems only to allow uncoloured systems to exist free of one another;
colour is confined in clusters where the net colour cancels out.

GLUONS

Combining electrostatics with relativity and quantum theory generated the successful
theory known as quantum electrodynamics (QED). The idea behind the quantum
chromodynamic (QCD) theory of quark forces is that colour generates them in the

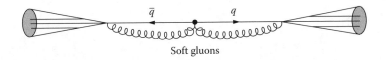

$\bar{q}$ q

Soft gluons

(a) Two jets of hadrons produced, one on each side. (1975 on)

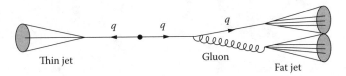

Thin jet   Gluon

Fat jet

QCD: Quark emits gluons. If there is a small angle between quark and
gluon then their separate jets are not resolved. Thin jet on one side,
fat jet on other side. (TASSO group, 1979)

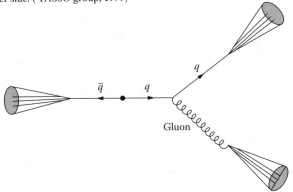

(b) QCD: If there is a larger angle between quark and gluon there are three
distinct jets, one each from quark, antiquark, and gluon. [(1979) and LEP]

FIGURE 7.4 Quarks and gluons in electron-positron annihilation.

same way that electric charge generates electromagnetic forces. Mathematically, as
quantum electrodynamics is to electric charge, so is quantum chromodynamics for
colour.

In quantum *electro*dynamics, photons necessarily exist: they are the massless,
spin 1 carriers of the electromagnetic force between electrically charged objects. In
quantum *chromo*dynamics, the analogous requirement is the existence of 'gluons:'
massless, spin 1 carriers of the force between coloured objects.

In quantum *electro*dynamics, the acceleration of an electric charge leads to radiation of photons. Analogously in quantum *chromo*dynamics, the acceleration of colour
radiates gluons (Figure 7.4). Both of these phenomena occur in the electron-positron
annihilation process if quarks are produced. The electron (e^-) and the positron
(e^+) annihilate and produce a photon (γ) which then converts into a quark (q) and

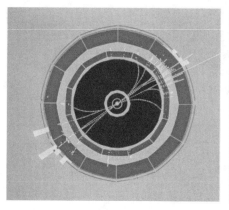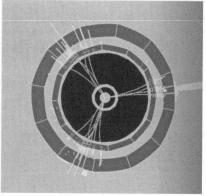

FIGURE 7.5 Jets at LEP. These computer displays show the 'beam's-eye view' of the cylindrically symmetric detector layers within the ALEPH experiment at CERN's LEP collider. An electron and positron have annihilated along the axis perpendicular to the page. In the left image, a quark and antiquark have emerged, which fly off back to back conserving momentum, and have immediately seeded new particles, hadrons, that form the two jets seen in the detector (compare with Figure 7.4a). In the image on the right, an additional jet of particles appears, and none of the jets is back to back. In this case either a quark or an antiquark has radiated a gluon through the strong force. The gluon, like the quark and antiquark, has emerged into the detector as a jet of hadrons, giving three jets in all (compare with Figure 7.4b).

antiquark ($\bar{q}$). This sequence of events is conventionally written:

$$e^+ + e^- \rightarrow \gamma \rightarrow q + \bar{q}$$

But the quark and antiquark carry both colour and electrical charge and, in the act of being produced, they radiate gluons and photons. So the real process is:

$$e^- + e^+ \rightarrow \gamma \rightarrow (q + \text{photons} + \text{gluons}) + (\bar{q} + \text{photons} + \text{gluons})$$

From established QED one can calculate how much of what is observed is due to photon radiation. You can then study what is left over, seek characteristics associated with gluon radiation from coloured quarks, and compare the resulting phenomena with QCD predictions.

It would be easy to test the theory if we could directly detect the quarks and gluons created by the electron–positron annihilation. However, Nature is not so kind, as only conventional pions, protons, and other clusters of quarks or gluons appear; isolated quarks and gluons do not emerge. QCD predicts the following scenario and experiment is consistent with it. Immediately after their production, the quark and antiquark move off in opposite directions. Initially they feel no force, but as they separate, the energy in the force field between them grows, eventually becoming so great that its energy "E" exceeds the "mc^2" needed to create further quarks and antiquarks. The initial quark and antiquark cluster together with these spontaneously created ones so quickly, forming mesons and baryons, that the original quark and antiquark are not detected; instead, two oppositely directed showers of hadrons emerge. The quark and antiquark are long gone; the two jets of hadrons are all that remain, indicating where the original basic particles once were. By studying these jets, the properties of the quarks or gluons that seeded them can be deduced.

QCD predicts that at high energies the quark and antiquark usually carry off most of the energy with the glue collimated along their direction of motion, carrying little energy itself. In such circumstances, two distinct jets of particles emerge.

Experiments with electron and positron beams have covered a vast range of energies. These include relatively low-energy collisions studied in the 1970s, where the beams had only 1 or 2 GeV energy apiece, to collisions at 100 GeV and beyond at CERN (LEP) in the 1990s. In the experiments at low energies, the jets are smeared out, distributed about the direction of the parent quark's motion. QCD predicts that at higher energies the jets should become increasingly collimated and the data confirm this. Although these 'two-jet' events dominate the data, there is a chance that the quark (or antiquark) radiates a gluon which carries of a lot of momentum, and deflects the quark (antiquark) from its path (Figure 7.4b). If the angle between the gluon and quark is small, then it will not be possible to distinguish the hadrons coming from each; there will be a thin jet on one side and a fat jet on the other. Sometimes the deflection of the quark when it emits a gluon is large enough that the individual jets can be identified. This will yield three jets of particles (Figure 7.4b).

Millions of examples of such"three jet" events have been seen. The way that the energy is shared among the jets, and the relative orientation of the jets in space, show that two are emerging from spin 1/2 quarks while the third originates in a spin 1 object — the gluon — as predicted by quantum chromodynamics.

QUANTUM CHROMODYNAMICS AND ELECTRODYNAMICS

As QCD is mathematically so similar to QED, then we might expect to find similar behaviours for the forces between quarks in clusters (qqq like the proton or $q\bar{q}$ like the pion) and the electromagnetic forces between the electrons and nuclei of atoms. Such similarities are indeed observed, most noticeably in the hyperfine splittings between certain energy levels of the atom or quark clusters. In hydrogen there is a magnetic interaction between the electron and proton which contributes a positive or negative amount to the total energy, depending upon whether the electron and proton are spinning parallel or antiparallel (total spin 1 or 0). In quark clusters there is an analogous 'chromomagnetic' interaction between pairs of quarks which adds to the energy of a spin 1 pair, and depletes that of a spin 0 pair. Such a splitting is indeed seen for quark–antiquark systems (mesons), for example, where the $^3S_1$ combinations (ρ, K^*, ϕ) are some hundreds of MeV heavier than their $^1S_0$ counterparts (π, K, η).

Not only does this behaviour illustrate the similarity between QCD for quark clusters and QED for atoms, but it also shows the relative strengths of the two forces. In atoms these hyperfine splittings are of the order of an electron-volt in energy, about

100 million times smaller than the effect in quark clusters. Most of this is due to the fact that atoms are between one and ten million times larger than quark clusters, which implies that the energy splittings should be one to ten million times smaller on dimensional grounds. That they are yet smaller by a factor of ten to a hundred is because the electromagnetic force in atoms is intrinsically that much weaker than the quark forces in hadrons.

When discussing the intrinsic strength of the inter-quark forces and comparing QCD with QED, it is important to specify the distances involved over which the colour force is acting. Although QED and QCD are mathematically almost identical, the replacement of *one* (electric) charge by *three* colours causes the two forces to spread out spatially in totally different ways (Figure 7.6). QCD predicts that when

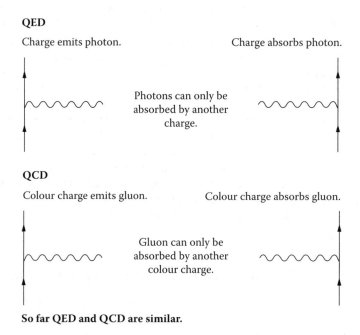

QED

Charge emits photon.   Charge absorbs photon.

Photons can only be
absorbed by another
charge.

QCD

Colour charge emits gluon.   Colour charge absorbs gluon.

Gluon can only be
absorbed by another
colour charge.

So far QED and QCD are similar.

The difference
In QCD the gluons also carry colour charge. This gives a new possibility:

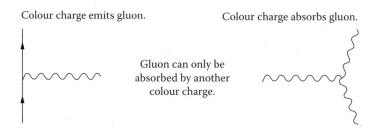

Colour charge emits gluon.   Colour charge absorbs gluon.

Gluon can only be
absorbed by another
colour charge.

FIGURE 7.6 QED and QCD: similar theories with far-reaching differences. (*Continued.*)

So the electromagnetic force between electrons

and the colour force between quarks

are different because a gluon can split into two gluons on the journey (QCD),

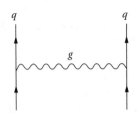

whereas a photon cannot split into two photons (QED) as they carry no electrical charge. It turns out that this causes the electromagnetic and interquark forces to behave quite differently. In particular, the interquark force becomes weaker at short distances and strong at large distances, in agreement with data and explaining the paradox on p. 88.

FIGURE 7.6 (Continued).

coloured objects like quarks are much closer to one another than a fermi (10^{-15} m), the forces between them are almost nonexistent; then, as the quarks move apart, the energy in the force field between them grows. According to the theory, it would take an infinite amount of energy to separate the quarks by more than a fermi. Thus it is impossible to separate an individual quark from the neighbourhood of its companions. The consequence is that quarks are permanently confined in clusters — the hadrons. Protons, pions, and similar particles all have sizes of the order of 1 fermi, which is a consequence of this behaviour.

At high energy: Free quarks in the proton but no escape

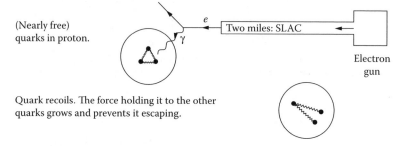

(Nearly free) quarks in proton.

Two miles: SLAC

Electron gun

Quark recoils. The force holding it to the other quarks grows and prevents it escaping.

The strong force isn't always strong

The picture of the proton changes with the energy of the photons that see it.

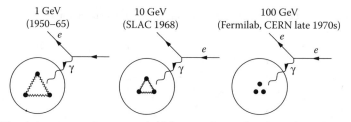

1 GeV (1950–65)	10 GeV (SLAC 1968)	100 GeV (Fermilab, CERN late 1970s)

The photon sees quarks tightly bound. Genuine strong force, complicated to deal with mathematically. 1950–70, limited to low energies: this tight binding slowed up theoretical progress.

At higher energies quarks appear almost free. Force no longer strong. QCD explains this and predicts that the force gets weaker still as energy increases. 1970 experiments at CERN seem to confirm this behaviour.

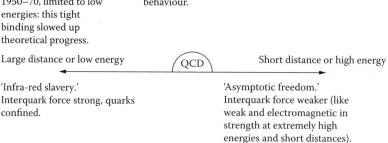

Large distance or low energy   QCD   Short distance or high energy

'Infra-red slavery.'
Interquark force strong, quarks confined.

'Asymptotic freedom.'
Interquark force weaker (like weak and electromagnetic in strength at extremely high energies and short distances).

FIGURE 7.7 QCD solves the quark force paradox.

The original SLAC experiments (p. 88) discovered these very phenomena — high-energy electrons scattered from quarks that were apparently free and yet stayed confined in clusters (Figure 7.7). QCD theory explains this, has survived 40 years of detailed examination and been tested to accuracies of better than one part in a thousand.

Before proceeding, let me answer a question that might have occurred to you. Does the notion of colour and its role in generating forces among the quarks mean that we now have five fundamental forces — gravity, weak, electromagnetic, strong nuclear force and this new quark (colour) force?

In fact, we still have only four: the strong nuclear force between neutrons and protons is now recognised to be a complicated manifestation of the more fundamental colour force acting between their constituents — the quarks. It may be helpful to draw a historical analogy. In the early 19th century inter-molecular forces were thought to be fundamental. We now realise that they are but complicated manifestations of the more fundamental electromagnetic force acting on the atomic constituents — the electrons.

There is a profound parallel between

electric and **colour** forces
acting on
    electrons and **quarks**
which are the constituents of
    atoms and **protons/neutrons**.
These in turn form
    molecules and **nuclei.**
The historically identified
    molecular and **strong nuclear** forces
are manifestations of the more fundamental
    electric and **colour** forces
acting on the constituent
    electrons and **quarks**

Nature does indeed appear to be efficient, not just in the fundamental particles (leptons and quarks) but in the forces that bind them to form bulk matter. It turns out that it is even more profound. These four forces are actually only three: the electromagnetic and weak forces are two manifestations of the more fundamental "electroweak" force. The discovery of this unification of forces forms the next part of our story.

8 The Electroweak Force

HISTORY

The weak force can change particles from one flavour to another. The most familiar example is when it changes a neutron into a proton, as in the form of radioactivity known as β decay (beta decay). Such a transmutation was responsible for Becquerel's 1896 discovery of the beta particles (electrons) produced when uranium nuclei decay.

It is the neutrons in nuclei that are the source of beta radioactivity:

$$n^0 \rightarrow p^+ + e^- + \bar{\nu}$$

(The e^-, electron, is the beta particle, and $\bar{\nu}$ an antineutrino, successfully predicted by Pauli in 1931 to explain the apparent imbalance of energy and momentum in such processes.) In some nuclei a proton can turn into a neutron and emit a positron — the positively charged antiparticle of an electron — by a similar process: $p^+ \rightarrow n^0 + e^+ + \nu$ (Chapter 3).

As neutrons and protons form the nuclear seeds of the atomic elements, such transmutations cause one element to change into another. The weak force is thus an alchemist. More profound insights emerged in the final quarter of the 20th century once it had become apparent that neutrons and protons are, in their turn, clusters of quarks:

$$n^0(d^{-1/3}; d^{-1/3}; u^{+2/3})$$
$$p^+(u^{+2/3}; d^{-1/3}; u^{+2/3})$$

(The superscripts remind us that up and down quarks have electrical charges that are fractions 2/3 and $-1/3$ of a proton's charge. The combination ddu has total of zero as required in a neutron.) The two clusters differ in that replacing one 'down' quark in the neutron by an 'up' quark yields the cluster that we call a proton. The fundamental cause of beta-radioactivity is the quark decay (Figure 8.1):

$$d^{-1/3} \rightarrow u^{+2/3} + e^- + \bar{\nu}.$$

Bury this down quark in a neutron and its beta-decay causes that neutron to decay:

$$n^0(d^{-1/3}d^{-1/3}u^{+2/3}) \rightarrow p^+(u^{+2/3}d^{-1/3}u^{+2/3}) + e^- + \bar{\nu}.$$

Elegant properties of the weak interaction appear in quark decay which are masked in neutron (let alone nuclear) decays. For example, the weak transmutation of a down to an up quark has identical properties to that of an electron into a neutrino and to that of a muon turning into another neutrino — called the 'muon-neutrino' (denoted ν_μ) to distinguish it from the former 'electron-neutrino' (denoted ν_e).

It was in 1933, more than 30 years before quarks were thought of, that Enrico Fermi proposed the first theory of beta decay. Fermi believed neutrons and protons to be structureless building blocks of matter and assumed as much in attempting to

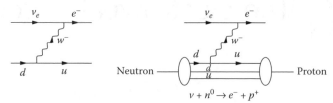

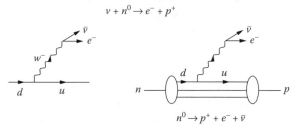

A neutrino and a down quark interacting through the weak interaction produce an electron and an up quark. Bury the down quark in a neutron (two down quarks and one up quark) and this becomes the process

$$v + n^0 \rightarrow e^- + p^+$$

Twist the figure around and we see the weak decay of a down quark:

$$d \rightarrow u + e^- + \bar{v}$$

Buried in the neutron this yields the weak decay of a neutron:

$$n^0 \rightarrow p^+ + e^- + \bar{v}$$

FIGURE 8.1 Quark weak interactions and neutron decay.

understand *neutron* beta decay and in making his embryonic theory of the weak force. Although superceded today, his line of attack was essentially correct. The following story illustrates the progression from conception to birth of a fully fledged testable theory of the weak force that combines it with electromagnetism.

Fermi's theory of beta decay was inspired by quantum electrodynamics where a neutron or proton absorbs a photon at a single point in space-time and preserves its electrical charge. Fermi proposed that something analogous occurred in β decay: in his theory, the change of charge in the decay of the neutron into a proton is caused by the emission of an electron and an antineutrino at a point.

The electron and antineutrino each have spin 1/2 and so their combination can have spin total 0 or 1. The photon, by contrast, has spin 1. By analogy with electromagnetism, Fermi had (correctly) supposed that only the spin 1 combination emerged in the weak decay. To further the analogy, in 1938, Oscar Klein suggested that a spin 1 particle ('W boson') mediated the decay, this boson playing a role in weak interactions like that of the photon in the electromagnetic case (Figure 8.2).

In 1957, Julian Schwinger extended these ideas and attempted to build a unified model of weak and electromagnetic forces by taking Klein's model and exploiting an analogy between it and Yukawa's model of nuclear forces. As the π^+, π^- and π^0 are

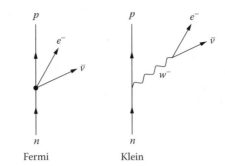

The Fermi and Klein models of β decay.

FIGURE 8.2 Weak decay.

exchanged between interacting particles in Yukawa's model of the nuclear force, so might the W^+, W^-, and γ be in the weak and electromagnetic forces.

However, the analogy is not perfect. The strong nuclear forces act independent of the electrical charges of the particles involved (π^+, π^-, and π^0 exchanges give comparable strengths) whereas the weak and electromagnetic forces are very sensitive to electrical charge: the forces mediated by W^+ and W^- appear to be more feeble than the electromagnetic force. Schwinger suggested that this difference in strength between electromagnetic and "weak" is subtle and to some extent illusory. When we think of the weak force as feeble, it is because we are describing it in the Fermi way where the interaction occurs at a single point in space and time. The Klein and Schwinger picture is different from this; in their theory the interaction occurs between distant particles much as the electromagnetic interaction does, the photon messenger being replaced by a W particle. The apparent strength of the weak force depends not only on the strength of the W's interaction with matter but also on its mass: the more massive the W, the more feeble would the interaction appear to be (see also Figure 4.3).

If the W boson links to the neutron–proton transition and also to the electron–neutrino transition with electromagnetic strengths, then its mass would have to be somewhere in the range of 10 to 100 GeV if the observed eponymous "weak" strength was to be explained. Refinements in the theory and in experimental precision in the 1960–1970s implied that the mass of these W bosons should be about 82 GeV. Experiments capable of producing such massive objects began at CERN in 1982, and on 21 January 1983 their discovery was announced — their mass precisely as predicted.

The search and discovery of the W is described on p. 119. On a first read, you could profitably go there directly. The intervening pages give a more detailed description of some of the ideas in what has become known as the "electroweak" theory, and of the experiments that culminated in the discovery of the W boson.

A NEW FORM OF WEAK INTERACTION IS PREDICTED

In quantum electrodynamics a photon can be emitted or absorbed by an electrically charged particle. The amount of charge that a particle has determines the likelihood of it interacting electromagnetically: the greater the charge, the bigger the chance. The charges of electron, proton, and uranium nucleus are proportional to the *numbers* $-1, +1, +92$. Refer to Table 8.1.

 If neutrinos are fired at neutrons, they convert into electrons and the neutron converts into a proton:

$$\nu^0 + n^0 \rightarrow e^- + p^+$$

which is similar to the electromagnetic process:

$$e^- + p^+ \rightarrow e^- + p^+$$

(I have put their electrical charges as superscripts so that it is easy to see how the charges balance in the two cases.) The basic idea is that the two processes are essentially the same but for the way the charges flow. The Feynman diagrams illustrate how the photon and W boson play corresponding roles. The following weak interaction can occur instead of electromagnetic scattering in an electron–proton collision:

$$e^- + p^+ \rightarrow \nu^0 + n^0$$

in which case a W^- is exchanged.

Table 8.1 The Puzzle of the Electron and Proton Electrical Charge: Episode 1

The electrical charges of electron and proton are opposite in sign and exactly equal in magnitude. This is the reason that atoms with the same number of electrons and protons have a net charge of zero, a fact that is so familiar that we don't give much thought to it. Yet this is really an astonishing phenomenon. Leptons and nuclear matter have appeared to be totally independent of each other in everything that we have discussed so far and yet the charged leptons and protons have precisely the same magnitudes of charge.

The simplest explanation is to suppose that electrical charge is some sort of external agent, a property of space perhaps, that is attached to matter in discrete amounts. Thus if we start with electrically neutral matter (neutron and neutrino), then the addition of the unit of charge to the neutron or removal from the neutrino will yield charged particles with exactly balanced amounts of charge. These are the proton and electron.

This was the sort of idea in Fermi's mind in 1933. The equality of charges is rationalized by relating the electron and proton to neutral partners, the neutrino and neutron, respectively. Thus we see the first emergence of the idea of families; two leptons (neutrino and electron) and two particles of nuclear matter (neutron and proton).

The idea of families persists today but with down and up quark replacing the neutron and proton family. However, the idea of adding charge to fundamental neutral matter has been lost because the neutron is now known to be built from quarks which are themselves charged. The equality of electron and proton charges is therefore resurrected as a fundamental puzzle, suggesting that there exists a profound relationship between quarks and leptons.

Just as the electron–photon coupling is described by a number (Table 4.4b), so could we denote the $\nu^0 \rightarrow e^- W^+$ and the $e^- \rightarrow \nu^0 W^-$ interactions by numbers. However, that would be closing our eyes to an obvious further symmetry in nature: these processes are not independent of one another; in fact, they have the same strength. How can we build this symmetry into our theory at the outset?

To do so we describe the electron and neutrino by a single entity, a matrix (see Table 8.2) with two members:

$$\begin{pmatrix} \text{Chance of being neutrino} \\ \text{Chance of being electron} \end{pmatrix}$$

so that $\begin{pmatrix} 1 \\ 0 \end{pmatrix}$ represents a neutrino, and $\begin{pmatrix} 0 \\ 1 \end{pmatrix}$ represents an electron. The W^+ and W^- are then represented by the following 2×2 matrices:

$$W^+ = \begin{pmatrix} 0 & 1 \\ 0 & 0 \end{pmatrix}; \quad W^- = \begin{pmatrix} 0 & 0 \\ 1 & 0 \end{pmatrix}$$

Table 8.2 Matrices

Many phenomena require more than just real numbers to describe them mathematically. One such generalisation of numbers is known as "matrices." These involve numbers arranged in columns or rows with their own rules for addition and multiplication. Addition has no surprises:

$$\begin{pmatrix} a & b \\ c & d \end{pmatrix} + \begin{pmatrix} A & B \\ C & D \end{pmatrix} = \begin{pmatrix} a+A & b+B \\ c+C & d+D \end{pmatrix}$$

but multiplication is less obvious: it involves the product of all elements of intersecting rows and columns:

$$\begin{pmatrix} a & b \\ c & d \end{pmatrix} \times \begin{pmatrix} A & B \\ C & D \end{pmatrix} = \begin{pmatrix} aA+bC & aB+bD \\ cA+dC & cB+dD \end{pmatrix}$$

The above matrices are 2×2 — two rows and two columns — and from this generalised perspective, conventional numbers are 1×1 matrices! (or less trivially $N \times N$ matrices with the north-west to south-east diagonal elements all identical and all other entries zero).

Matrices can have any number of rows and any number of columns; they do not have to be the same. Thus we can have column matrices such as

$$\begin{pmatrix} A \\ C \end{pmatrix} = A \begin{pmatrix} 1 \\ 0 \end{pmatrix} + C \begin{pmatrix} 0 \\ 1 \end{pmatrix}$$

When a 2×2 matrix multiplies such a column matrix, the result is the same as above with B and D thrown away.

$$\begin{pmatrix} a & b \\ c & d \end{pmatrix} \times \begin{pmatrix} A \\ C \end{pmatrix} = \begin{pmatrix} aA+bC \\ cA+dC \end{pmatrix}$$

The theory then requires that interactions of common strength among particles are described by multiplying the matrices together.

Using the multiplication rules shown in Table 8.2, we can first of all check that these matrices do indeed faithfully represent the pattern of interactions observed:

$$\begin{pmatrix} 0 & 1 \\ 0 & 0 \end{pmatrix} \times \begin{pmatrix} 0 \\ 1 \end{pmatrix} = \begin{pmatrix} 1 \\ 0 \end{pmatrix}$$

represents

$$W^+ + e^- = v^0 \quad \checkmark$$

while

$$\begin{pmatrix} 0 & 0 \\ 1 & 0 \end{pmatrix} \times \begin{pmatrix} 1 \\ 0 \end{pmatrix} = \begin{pmatrix} 0 \\ 1 \end{pmatrix}$$

represents

$$W^- + v^0 = e^- \quad \checkmark$$

These two processes have the same strength. Furthermore:

$$\begin{pmatrix} 0 & 1 \\ 0 & 0 \end{pmatrix} \times \begin{pmatrix} 1 \\ 0 \end{pmatrix} = 0$$

implies that $W^+ + v^0$ does not happen. Nor do W^- and e^- interact together, in agreement with the matrix multiplication

$$\begin{pmatrix} 0 & 0 \\ 1 & 0 \end{pmatrix} \times \begin{pmatrix} 0 \\ 1 \end{pmatrix} = 0$$

These matrix multiplications match one-on-one with the set of processes that are observed and so appear to be the mathematics needed to describe the weak interaction. Indeed, if one attempts to build a theory of the weak interaction by imitating the successful quantum electrodynamic theory of electromagnetism, one is inexorably led to the introduction of these matrices and that all possible multiplications represent physical processes. This is an important result: there are possible multiplications that we have not considered so far.

For example, see what happens if we multiply the matrices for W^- and W^+: the resulting matrix will correspond to a particle produced in a $W^- W^+$ collision:

$$\begin{pmatrix} 0 & 1 \\ 0 & 0 \end{pmatrix} \times \begin{pmatrix} 0 & 0 \\ 1 & 0 \end{pmatrix} = \begin{pmatrix} 1 & 0 \\ 0 & 0 \end{pmatrix}$$

$$W^+ \quad + \quad W^- \quad = \quad ?$$

This produces a matrix that we have not met before and so implies that a new particle exists. This will be a partner to the W^+ and W^- and has no electrical charge (it was formed by a W^+ and W^- interacting). You might be tempted to guess that this is the photon, but this is not so, as can be seen by studying its interaction with a neutrino:

$$\begin{pmatrix} 1 & 0 \\ 0 & 0 \end{pmatrix} \times \begin{pmatrix} 1 \\ 0 \end{pmatrix} = \begin{pmatrix} 1 \\ 0 \end{pmatrix}$$

$$? \quad + \quad v^0 \quad = \quad v^0$$

The matrices show that this new particle interacts with the electrically neutral neutrino, whereas photons do not. Thus the Schwinger model was on the right track but the neutral partner of the W^+ and W^- is not simply the photon: the four matrices

$$\begin{pmatrix} 1 & 0 \\ 0 & 0 \end{pmatrix}, \begin{pmatrix} 0 & 1 \\ 0 & 0 \end{pmatrix}, \begin{pmatrix} 0 & 0 \\ 1 & 0 \end{pmatrix}, \begin{pmatrix} 0 & 0 \\ 0 & 1 \end{pmatrix},$$

correspond to four particles Z^0, W^+, W^-, γ (see also Table 8.3 for more about this). Consequently, a new form of weak interaction is predicted when the Z^0 is involved.

Table 8.3 SU(2)

The astute reader may have noticed that the collision of a W^+ and W^- could be represented by either $W^+ + W^-$ whose matrix representation is:

$$\begin{pmatrix} 0 & 1 \\ 0 & 0 \end{pmatrix} \times \begin{pmatrix} 0 & 0 \\ 1 & 0 \end{pmatrix} = \begin{pmatrix} 1 & 0 \\ 0 & 0 \end{pmatrix}$$

or by $W^- + W^+$ in which case it would be:

$$\begin{pmatrix} 0 & 0 \\ 1 & 0 \end{pmatrix} \times \begin{pmatrix} 0 & 1 \\ 0 & 0 \end{pmatrix} = \begin{pmatrix} 0 & 0 \\ 0 & 1 \end{pmatrix}$$

which begs the question of which of these represents the particle produced in the collision.

Deeper aspects of the theory than we can go into here require that the 2×2 matrices must be restricted to those whose top-left to bottom-right diagonal numbers add to zero ("traceless matrices"). These are:

$$W^+ = \begin{pmatrix} 0 & 1 \\ 0 & 0 \end{pmatrix}; \quad W^- = \begin{pmatrix} 0 & 0 \\ 1 & 0 \end{pmatrix}; \quad W^0 = \begin{pmatrix} 1 & 0 \\ 0 & -1 \end{pmatrix}$$

which implies that W^0 is represented by the difference of the two possibilities above. These 2×2 traceless matrices are known as SU(2) matrices in mathematical jargon. The analogous $N \times N$ traceless matrices are called SU(N) matrices. Just as SU(2) contains $3 = 2^2 - 1$ (the W^+, W^-, and W^0), so does SU(3) have $8 = 3^2 - 1$ (which are the eight coloured gluons).

The sum of the two matrices is $\begin{pmatrix} 1 & 0 \\ 0 & 1 \end{pmatrix}$, which is known as a U(1) matrix. The fact that the collision of a W^+ and W^- could produce a photon instead of the W^0 is related to the two independent matrices arising from a $W^+ W^-$ collision. The total theory of weak (W) and electromagnetic (γ) interactions is mathematically an SU(2) $\times$ U(1) theory.

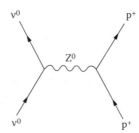

The old familiar processes are known as 'charged' weak interactions (alluding to the role of the charged W^+ and W^-), and the new process is called a weak 'neutral'

interaction, or in the jargon: a 'neutral current' interaction. The Z^0 can be exchanged between a neutrino and a proton and cause the previously unobserved process $v^0 + p \rightarrow v^0 + p$ to occur. Thus neutrinos can interact with matter without swapping the charges around.

THE DISCOVERY OF NEUTRAL CURRENTS

In "charged current" interactions, a neutrino converts into a charged particle, for example, an electron or muon, which can be detected easily by virtue of its electrical charge. Electrically neutral particles, on the other hand, are notoriously difficult to detect: indeed we may recall that 25 years elapsed between Pauli proposing that neutrinos exist and Cowan and Reines detecting them. Furthermore, they did so by observing the charged electrons that were produced as a result of the neutrinos interacting with matter. Small wonder that no one had ever seen processes where a neutrino comes in, scatters, and remains a neutrino. However, if the model was correct, such 'neutral current interactions' must occur.

One way that they would show up would be when a neutrino hit an electron or proton in an atom, bounced off it and, in doing so, set the charged particle into motion. Experimentalists searched their data, looking for examples where a charged particle, such as an electron or proton, suddenly moved off when neutrinos were fired at them, and where no other visible phenomena accompanied the interaction. By painstaking effort, evidence for such ephemeral processes was obtained in 1973 at CERN by the Gargamelle collaboration. (Thirty years later, such neutral current processes have become part of of the high-energy particle physicists' toolkit. They have even been used to measure what is going on in the heart of the Sun; Chapter 12.)

By calculating the Feynman diagrams for this process, the implications of the theory can be worked out. Thus were predicted various properties of the interaction (how much energy the proton tends to absorb, what direction it tends to recoil in, whether it remains a proton or breaks up, and so on). It also made clear predcitions for what would happen in experiments with electron targets instead of protons.

The discovery of these previously unknown forms of interaction was the first good evidence for the validity of the electroweak theory. All of these processes were observed and agreed with the predictions of the theory if the Z^0 and W^+, W^- have masses of about 90 and 80 GeV, respectively. By 1979, the quantitative successes of the theory led physicists to accept it as (almost) proved to be correct, even though the crucial production of W^+, W^- and Z^0 bosons in the laboratory was still awaited.

ELECTROMAGNETIC AND WEAK INTERACTIONS
GET TOGETHER

I have glossed over some points of detail in the description of the weak interaction so far. The essential features have been correctly described but the following additional remarks are needed to show how the weak and electromagnetic interactions are wedded.

Recall that it was the fact that the weak interaction caused $v_e \rightleftharpoons e^-$ or $v_\mu \rightleftharpoons \mu$ for leptons, and $u \rightleftharpoons d$ for quarks that led to the introduction of matrices. Specifically, the electron and its neutrino (or the muon and its neutrino or the up and down quarks) are supposed to form doublets of 'weak isospin:'

$$Q \equiv \begin{pmatrix} u \\ d \end{pmatrix}, \quad L_1 \equiv \begin{pmatrix} \nu_e \\ e^- \end{pmatrix}, \quad L_2 \equiv \begin{pmatrix} \nu_\mu \\ \mu^- \end{pmatrix}$$

In 1953 Gell-Mann and Nishijima had noted, in a different context, that the electric charges of particles in isospin doublets are in general given by:

$$Charge = \pm\frac{1}{2} + \frac{Y}{2}$$

where $\pm 1/2$ is for the upper or lower members of the pair while Y is an additive quantum number called 'hypercharge.' Leptons have different electrical charges from quarks and this is a result of the lepton doublets $L_{1,2}$ having $Y = -1$ (hence charge $= 0$ or -1 as observed), whereas the quark doublet has $Y = 1/3$ (hence charges 2/3 and $-1/3$). The weak interaction model described so far has taken no account of this hypercharge degree of freedom nor that quarks and leptons have different values of it. Sheldon Glashow in 1961 was the first to do so, and produced the first viable model that brought the weak and electromagnetic interactions together (for which he shared the Nobel Prize in 1979).

The matrix mathematics that described the weak isospin is known as 'SU(2)' (the 2 refers to the doublet nature of the Q, L_1, L_2 above, or of the 2×2 representations of the W particles; SU is a mathematical classification, see Table 8.3). Thus the matrix theory of the weak interaction described so far is called an SU(2) theory.

The hypercharge on the other hand is a real number. In matrix jargon we think of it as a 1×1 matrix and so the mathematics involving such numbers is called U(1), the analogue of SU(2). Combining the weak isospin and hypercharge yields an SU(2) $\times$ U(1) theory.

The new feature that enters, as compared with the previous treatment, is that in addition to the W^+, W^-, W^0 of the SU(2) piece (which couple to matter with a common strength g_2), there is a fourth particle from the U(1) piece, an electrically neutral 'B^0' (which couples to matter with a strength g_1). The relative strengths g_1 and g_2 are quite arbitrary and are conventionally described in terms of a parameter θ_W (known ironically as the Weinberg angle even though this model was first constructed by Glashow in 1961).

Now we are almost home and dry. The physical photon couples to electric charge and is a quantum superposition of W^0 and B^0. The W^+ and W^- transport the 'charged' weak interaction known for 50 years. In addition, there is a new 'neutral' weak interaction carried by Z^0, the orthogonal superposition of W^0 and B^0:

$$\text{SU(2)} \left. \begin{cases} W^+ \\ W^- \\ W^0 \end{cases} \right. \qquad \left. \begin{matrix} W^+ \\ W^- \end{matrix} \right\} \text{charged weak interaction}$$

$$\left. \begin{matrix} \times \\ \text{U(1)} \quad B^0 \end{matrix} \right\} \text{mix} \begin{cases} \gamma & \text{electromagnetism} \\ Z^0 & \text{neutral weak interaction (predicted)} \end{cases}$$

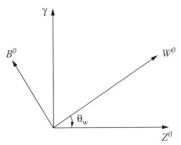

The physical γ and Z^0 are mixtures of the B^0 and W^0. If $\theta_w = 0$, then $W^0 \equiv Z^0$ and $\gamma \equiv B^0$, the weak and electromagnetic interactions would not be mixed together. Empirically $\sin^2 \theta_w \approx \frac{1}{5}$ and the photon is a superposition of both B^0 and W^0.

FIGURE 8.3 The Weinberg angle.

The way that this mixing occurs is beyond the scope of the present text. It is as if the W^0 and B^0 are represented by two axes in some abstract space. The γ and Z^0 are represented by two orthogonal vectors at some angle θ_W relative to these axes (Figure 8.3). This is the same θ_W that is related to the relative strengths g_1 and g_2.

There is one further subtlety that should be mentioned if you are wanting to go into all this more deeply. The weak interactions distinguish left from right (see Figure 4.4). For example, a neutrino spins only one way, known as left-handed, whereas an electron can be either left- or right-handed. Right-handed neutrinos do not exist in this "Standard Model," and the right-handed electron does not feel the weak force involving the W^+ or W^-. Technically it is the left-handed situations to which all of these matrices apply and the theory is known mathematically as $SU(2)_L \times U(1)$, the subscript denoting "left-handed." So although we talk loosely of having "unified" the weak and electromagnetic interactions, there are unresolved questions, not least why parity is violated in this idiosyncratic manner.

DOES ELECTRON SCATTERING RESPECT MIRROR SYMMETRY?

Not only can the Z^0 be exchanged by neutrinos interacting with matter, but it can also be exchanged by electrons. This can give rise to the following interaction:

$$e^- + p^+ \rightarrow e^- + p^+ :$$

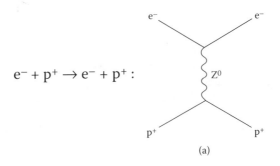

(a)

This is going to be swamped by the much more probable contribution from photon exchange:

$$e^- + p^+ \rightarrow e^- + p^+ :$$

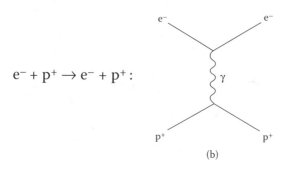

(b)

and so had never been seen in electron–proton scattering experiments. With the development of the electroweak theory, interest grew in the possibility of performing a very high-precision experiment to study electron scattering from protons to see if any evidence for this extra contribution could be found.

Charged weak interactions do not respect mirror symmetry. The Weinberg-Salam model uniting weak and electromagnetic interactions required also that there be no mirror symmetry in neutral current interactions, specifically that left-handed electrons interact with Z^0 more readily than do right-handed ones.

The 2-mile (3 km) long accelerator at Stanford, California, could produce high-energy electrons that corkscrew left-handed (as a neutrino) or right-handed. To test the theory, the experimentalists fired beams of these electrons at a target of protons and discovered that the left-handed electrons have a greater tendency to interact with the protons than do the right-handed ones. The left-handed excess was only one event in every 10,000 or so, but was a large effect within the sensitivity of the experiment.

Here was a new way of distinguishing the real world from the mirror world — there are both left-handed and right-handed electrons, but it is the former that prefer to interact in the real world. A new piece of evidence for a left-handed preference in nature had been obtained. The significance of this discovery was that the $SU(2)_L \times U(1)$ model combining weak and electromagnetic interactions had correctly predicted *both* that the left-handed electrons should win *and* by how much.

It is this observation of the left-handed excess that provided the first evidence that the weak neutral current couples to electrons, and the first *direct* proof that it does not respect mirror symmetry.

BOSONS

Glashow's model of weak and electromagnetic interactions was a matrix generalisation of standard electromagnetic theory. The relativistic quantum theory of the latter, quantum electrodynamics, had been developed several years earlier and implied that photons are massless. This is of course quite satisfactory because photons do indeed have this property, but is a problem if you try to imitate it in building a theory of the weak force: Glashow's model seemed to imply that photons and the W^+, W^-, and Z^0 were all massless. This would be a disaster, because even in the 1960s there was good evidence that the W and Z particles could not be light (a light W or Z would have been produced as easily as photons are, but none had been seen). So Glashow's model had no explanation of why W and Z had mass, whereas the photon was massless.

Stimulated by some developments in solid-state physics and quite independent of the above attempts to build theories of the weak interaction, in 1964 Peter Higgs in Britain and independently, Robert Brout and François Englert in Belgium, discovered that there was a loophole in the line of theoretical argument that required the photon to be massless. Higgs showed that it could have a mass if there also exists in Nature massive spinless particles (known as *Higgs bosons* after their inventor; see Chapter 13).

The discovery of the "Higgs mechanism" might seem to be rather academic because Nature does not appear to exploit it: empirically, photons are massless. But then Abdus Salam and Steven Weinberg independently suggested that Nature might make use of it for weak interactions. They showed that this could enable the photon to stay massless while W and Z gained quantifiable masses.

However, there was one remaining problem. While this created a theory that is satisfactory at low energies, no one was quite sure whether or not it made any sense when applied to high-energy interactions — it seemed to predict that certain processes would occur with infinite probability! The final link was provided by a young Dutchman, Gerhard 't Hooft in 1971. He proved that the theory is consistent and gives finite probabilities for physical processes at all energies.

If Higgs' mechanism is the way that W and Z particles become massive, then spinless Higgs particles ought to exist. Theoretical arguments, outlined in Chapter 13, suggest that the Higgs boson has a mass that is less than about 900 GeV. Experimental data in advance of the Large Hadron Collider (LHC) at CERN were even more restrictive and suggested that if Higgs' original idea is used by Nature in the simplest way, then the mass could be below 200 GeV. As we shall see, in the 40 years since he proposed his theory, it has been realised that there are other ways that Nature might exploit it. The LHC at CERN has been designed to cover the new energy frontier above 1000 GeV (1 TeV); see Chapter 13.

DISCOVERY OF THE W BOSON

At the risk of boring readers who ploughed through the previous ten pages, I shall first make a brief summary in case you have jumped to this page directly.

Schwinger's unified model involving W^+, W^-, and γ had promising features but also had some flaws. Most obvious of the problems was that W^+ and W^- both interact with neutrinos but photons do not, or at best only indirectly. A further neutral boson was required ("Z^0") which can directly couple to neutrinos as well as to other matter. This idea was first developed by Sheldon Glashow in 1961 and then extended by Steven Weinberg and Abdus Salam in 1967. They showed that the masses of Z^0,

W^+, and W^- could be mutually related, in particular the Z^0 cannot be lighter than the W^+ and W^- (and so *cannot* be the photon).

The electrically neutral weak interaction mediated by the Z^0 was discovered in 1973 and a subsequent series of tests led to the following predictions. If weak interactions are really intrinsically united with electromagnetism in this way, then their theory implied that:

$$m(W^+) = m(W^-) = 82 \pm 2 \text{ GeV}; m(Z^0) = 92 \pm 2 \text{ GeV}$$

(which means 82 and 92, respectively, with an uncertainty of 2 on either side). Compare these remarkable predictions with the actual values known today: $m(W) = 80.42 \pm 0.04$ GeV and $m(Z) = 91.188 \pm 0.002$ GeV. For their formulation of the theory, Glashow, Salam, and Weinberg shared the 1979 Nobel Prize for physics, even though neither W or Z had been detected at that time! Their theory had, after all, predicted a new phenomenon — the neutral weak interaction which was, in itself, a major step.

The task now was to produce the W and Z particles which are the quanta of weak radiation and bundles of weak energy. No machine existed powerful enough to do this.

CERN had just built a super proton synchrotron (SPS) capable of accelerating protons to energies of several hundred GeV. Then came the idea of creating antiprotons in a laboratory next to the SPS, accumulating them until huge quantities were available, and then injecting them into the SPS, a 4-mile-long circle of magnets where protons swung around in one direction and antiprotons in the other. When they met head-on, a cataclysmic annihilation of matter and antimatter occurred and, for a fraction of a second, conditions were similar to those present in the Big Bang. At these extremes, the W and Z particles were expected occasionally to appear.

The major difficulty was in storing the antiprotons. These are examples of antimatter, and as the machine, laboratory, physicists, and apparatus are all made of matter, it was essential to keep the antiprotons away from them and in a total vacuum. The enthusiasm of Carlo Rubbia, whose idea it was, and the brilliance of the machine designer, Simon van der Meer, combined to make the dream come true.

By the autumn of 1982 it was at last possible to smash the protons and antiprotons against each other. Usually they fragmented, their quarks and antiquarks creating showers of particles: one shower following the line of the proton, the other of the antiprotons. But occasionally one quark and one antiquark mutually annihilated and converted into radiant energy. The theory of Glashow, Salam, and Weinberg implied that at these huge energies where weak and electromagnetic interactions are united, these bundles of energy would be manifested as W or Z as frequently as photons. The W and Z are highly unstable and in less than 10^{-24} seconds have decayed into electrons, positrons, or neutrinos, which can shoot out in a direction perpendicular to that of the incident protons. Out of more than a million collisions, nine precious examples of such events had been found by January 1983.

The characteristics of those nine events were as conclusive a proof as a partial fingerprint can be at the scene of a crime. Their abundance suggested that the W is produced as easily as photons, and with a mass consistent with the theoretical expectations.

Collisions between protons and antiprotons, as at the SPS, produced W and Z at random. The proton or antiproton may have a specific amount of energy, but the quarks and antiquarks within shared this energy among themselves in such a way that the amount of energy carried by any particular quark or antiquark is not specified other than on the average. It was only when a quark and an antiquark collided with a combined energy that by chance equalled that needed to form a W or Z that these latter particles had a chance to appear. To make a Z^0 to order, the challenge was to use beams of electrons and their antiparticles, positrons. These could be accelerated to energies of some 45 GeV each, and their head-on collision totalling around 90 GeV would then be in the critical region needed to make a Z^0 when they annihilated. To accelerate the lightweight electrons and positrons to such energies, CERN built a 27 km ring — the Large Electron Positron Collider, LEP. (We shall see what this discovered in Chapter 10.)

CONCLUSIONS

In this chapter we have witnessed one of the great accomplishments of the latter half of the 20th century. Building a successful theory of weak interactions was in its own right a major feat, but the fact that it is so similar to, and subsumes, quantum electrodynamics is far more exciting. The electromagnetic force and the weak forces appear to have more to do with each other than either of them does with gravity or the strong nuclear force. This discovery gave further impetus to the belief that at even more extreme energies the strong nuclear force also will unite with the electroweak force.

In Chapter 7 we met the concept that any 'flavour' of quark could occur in three 'colours.' Just as the weak interaction theory emerged when we studied flavour doublets (i.e., electron and neutrino are two lepton flavours), one is tempted to ask what would happen if we played an analogous game with colour triplets, for example:

$$Q \equiv \begin{pmatrix} \text{Red quark} \\ \text{Yellow quark} \\ \text{Blue quark} \end{pmatrix}$$

and built an SU(3) theory involving 3×3 matrices.

The result is a theory similar to quantum electrodynamics but based on three colours. In place of one photon or three weak force carriers (W^+, W^-, Z^0), we now find eight colour carriers ('gluons'). This theory is quantum chromodynamics, whose properties were described in Chapter 7. Theories generated this way are called *gauge theories*. Electromagnetic and weak interactions are described by U(1) and SU(2) gauge theories; quantum chromodynamics is an SU(3) gauge theory.

Why Nature exploits the Higgs mechanism in the SU(2) case but not in the others, and why parity is violated when $W^\pm$ or Z^0 are involved but is satisfied when photons or gluons mediate the forces, are open questions. Apart from these asymmetries, the theories are very similar mathematically. Physicists suspect that this similarity is too marked to be an accident and believe that all of these forces are intimately related in some way. This has inspired attempts to develop Grand Unified Theories (GUTS for short) of all the forces and particles. The idea of GUTS builds on the symmetrical behaviours both of the forces, which, with their underlying $SU(3, 2, 1)$ mathematics, has become apparent, and the fundamental particles of matter, which at the start of the 1970s were still manifestly asymmetric. All this changed with some startling discoveries of new varieties of hadrons, and it was these that played a seminal role both in developing the electroweak theory and in turn stimulating the construction of Grand Unified Theories.

9 From Charm to Top

The theory of electromagnetic and weak interactions described in Chapter 8 was built upon the observation that leptons and quarks form pairs:

$$Q_1 \equiv \begin{pmatrix} u \\ d \end{pmatrix} \quad L_1 \equiv \begin{pmatrix} \nu_e \\ e^- \end{pmatrix}$$

$$L_2 \equiv \begin{pmatrix} \nu_\mu \\ \mu^- \end{pmatrix}$$

This is fine for the leptons and the up and down quarks, but leaves the strange quark in isolation and out of the weak interaction. Even looking at this, it seems asymmetric, almost demanding that there be another doublet of quarks to fill the empty space in the lower-left corner. In his original paper postulating quarks, Gell-Mann alluded to the possibility of a fourth quark (Figure 9.1) that formed a pair with the strange quark, 'by analogy with the leptons.' This idea was briefly pursued by Glashow, among others, but was then dropped because not a single hadron containing a 'charmed' quark was found.

Although the idea was not further pursued, it was not entirely forgotten. Glashow had been at Caltech in the early 1960s when Gell-Mann was developing his quark ideas, and it was during that period that Glashow had himself been developing his theory of weak interactions. He initially built it for leptons where it worked perfectly, successfully predicting the existence of neutral weak interactions, but it failed when applied to quarks. In addition to its successful prediction that neutral currents such as $\nu d \rightarrow \nu d$ can occur, it also predicted that down and strange quarks could transmute into one another: $\nu d \rightarrow \nu s$. The d and s quarks have the same electric charge and so the $\nu d \rightarrow \nu s$ process seemed unavoidable. However, such 'strangeness-changing neutral interactions' seemed not to occur in Nature. (If they had existed, the decays of a strange neutral kaon K^0 would include the possible processes $K^o \rightarrow \mu^+\mu^-$ or $K^0 \rightarrow e^+e^-$. However, none had been seen in over a million examples of K^0 decays. This contrasts with the decay of its electrically charged analogue, the K^+, where $K^+ \rightarrow \mu^+\nu$ is the most likely decay channel, occurring in nearly two thirds of all decays. Today with high-precision data, we know that if you study a billion K^0 decays, at most a handful of them will be the mode $K^0 \rightarrow \mu^+\mu^-$. This level of rarity agrees with the predictions based on the electroweak theory, including the charmed quark, which the following chapters will describe.)

It was realised that something must be preventing the neutral current from acting at full strength. Glashow mused that everything works fine for the leptons, the magic ingredient being that the leptons form *two* pairs. He revived the idea of the fourth quark and in 1970 with John Iliopoulos and Luciano Maiani showed that everything with the weak interaction theory would be perfect for both leptons *and* quarks if there

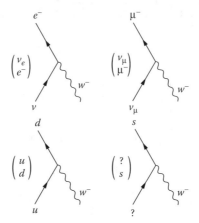

FIGURE 9.1 Leptons and quarks in weak interactions: the charmed quark. The properties of quark weak interactions are identical to those of leptons: as far as weak interactions are concerned, the up-down doublet and the electron–neutrino lepton doublet are identical. If there exists a fourth quark ("charmed quark") at '?,' the parallel would be very striking.

exists a charmed quark with charge +2/3, like the up quark, and if the quarks form two pairs similar to the leptons (Figure 9.2):

$$Q_1 \equiv \begin{pmatrix} u \\ d \end{pmatrix} \quad L_1 \equiv \begin{pmatrix} \nu_e \\ e^- \end{pmatrix}$$

$$Q_2 \equiv \begin{pmatrix} c \\ s \end{pmatrix} \quad L_2 \equiv \begin{pmatrix} \nu_\mu \\ \mu^- \end{pmatrix}$$

The unavoidable consequence of this was that there must exist scores of hadrons containing charmed quarks: so why had no one ever found any? If the theory was correct, the answer would be that the charmed quark is so massive that hadrons containing one or more of them would be too heavy to produce in the accelerators of the time. However, it was touch and go. If the GIM theory (pronounced 'Jim'), as it is known, was the solution to the absent strangeness changing neutral interactions, the charmed quark had to be heavy but not too heavy. Comparing the predictions of the theory and the constraints of the experiments, there appeared to be a small window of opportunity: charmed particles could possibly be light enough to be produced in experiments at the machines then available, when working at the extremes of their capabilities. Indeed, it was even possible that charmed particles had already been produced in experiments and not been recognised.

The best place to look seemed to be in electron–positron annihilation experiments (Table 9.1). If this annihilation occurred with sufficient energy, then a charmed quark (c) and a charmed antiquark (c̄) could be produced, and sometimes they might bind together making a meson built of a cc̄. During 1973 and 1974, some peculiar phenomena were seen when electron and positron annihilated at energies between 3 and

Table 9.1 Annihilating Matter and Antimatter

One of the most significant features that led to the discovery of charmed particles was the change of emphasis taking place in experimental high-energy physics during the 1970s.

Through the 1950s and 1960s, the central puzzle had been why so many strongly interacting particles existed. With the discovery of the Eightfold Way patterns and their explanation in terms of quark clusters, it was essential to establish if the spectroscopy of the states was indeed of this form.

To do this, protons were fired at stationary targets of matter, which were essentially protons and neutrons. The debris consisted mainly of strongly interacting particles that could be easily produced from them, namely, things made from up and down quarks. Strange quarks were light enough that pairs of strange quarks and antiquarks could be produced from the available energy in the collision and so strange mesons or baryons could also be produced and studied. Charmed quarks were very massive relative to the energies available in those days. As the everyday world is built from up and down flavours, a charmed quark and charmed antiquark have to be produced together and this was very difficult; hence, the failure to find charmed particles in those experiments.

Then in the late 1960s attention focussed on experiments involving leptons. One of the first was the classic experiment at Stanford where electron–proton collisions showed the first clear view of the quarks inside the proton (p. 105).

The most significant advances came with the development of 'storage rings' where electrons and their antiparticles (positrons) were stored and then collided head-on. Most frequently they scattered from one another, but occasionally the electron and positron, being matter and antimatter, annihilated each other, leaving pure energy. When enough energy is concentrated in a small region of space, new forms of matter and antimatter can be produced, in particular a quark and an antiquark.

Early machines were at Orsay near Paris and Frascati near Rome. These had enough energy to produce a strange quark and antiquark bound together as a ϕ meson. SPEAR at Stanford was the first able to make the spectroscopy of charm–anticharm pairs, the ψ states. Although the Υ (Upsilon) particle, made of a bottom quark and antiquark, was first produced at Fermilab using hadrons, its detailed study has come from electron–positron annihilations at Cornell, Stanford and KEK in Japan. The analogous mesons made of a top quark and antiquark are beyond the reach of electron–positron annihilation at present, and top flavours have only been produced in hadron experiments.

4 GeV. With hindsight, it is astonishing that no theorist gained a share of a Nobel Prize by suggesting that here, staring them in the face, was the evidence for charm.

As with the original quark idea, we have here another example of how scientific progress is not inexorably forward and in unison. Apart from a few committed aficianados, charm was not where most physicists were directing their attention. Everything changed dramatically on 10 November 1974 with the discovery of the

Leptons

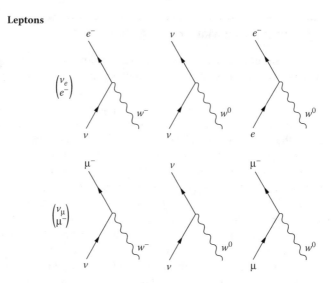

The leptons seem to exist in two separate pairs. No mixing occurs between the two pairs, i. e., transitions like

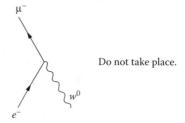

Do not take place.

FIGURE 9.2 The neutral current problem. (*Continued.*)

first example of a particle built from a charmed quark and a charmed antiquark — the J/ψ meson, nowadays referred to simply as the ψ. Not only was this the making of a scientific revolution, but there were great personal dramas too.

THE J/ψ MESON

The first machine with sufficient energy available for electron and positron to annihilate and produce the J/ψ meson was the storage ring 'SPEAR' at Stanford, California. (The electrons and positrons for this machine were provided by the two mile long electron accelerator in Figure 6.7.) It was here that a team led by Burton Richter discovered the J/ψ meson during the weekend of 9–10 November 1974. On Monday 11 November, the discovery was announced at a special meeting at Stanford. Samuel Ting from MIT was there. By a strange quirk of fate, he himself had already discovered this same particle in an experiment at Brookhaven Laboratory, New York. However, he had not yet announced it!

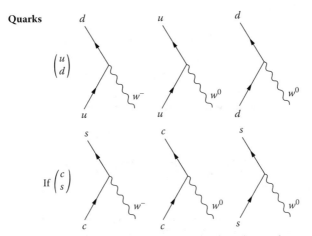

Quarks

The up and down quarks form a pair. If a charmed quark existed then a second pair would arise. Transitions like this:

There is good evidence that the transition $d \leftrightarrow s\ W^0$, does not occur in nature.

Such a transition is called a 'strangeness-changing neutral current:' 'neutral' because the down and strange quarks have the same electrical charge, and 'strangeness-changing' because they have different strangenesses. Such transitions are avoided if there are two 'generations' of quarks in analogous fashion to the way that two 'generations' of leptons do not admit the transition.

would not occur: The quarks and leptons would behave very similarly.

These pairings are now called 'generations' of quarks and leptons. (u, d) or (e^-, ν) form the first generation. (c, s) or (μ^-, ν) form the second generation.

(a)

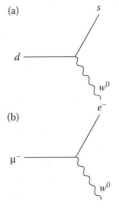

(b)

FIGURE 9.2 (Continued).

Ting had been leading a team studying collisions of protons and nuclei. They were not interested in the debris of pions that provided over 99% of the events; instead, they were seeking much rarer events where an electron and a positron were produced among the debris. By measuring the energies of the pair in each example, they could look to see if the total energy of the electron and positron tended to have some particular value rather than others. That is, what would happen if the collision had

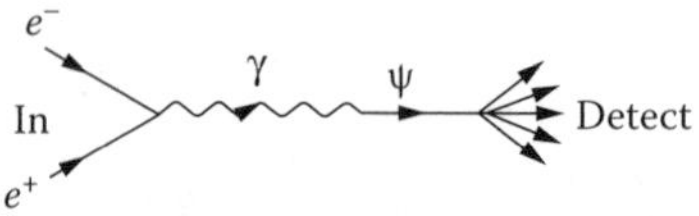

(a)

(a) Richter at SLAC produced it by annihilating electrons and positrons. He then detected its decay into hadrons.

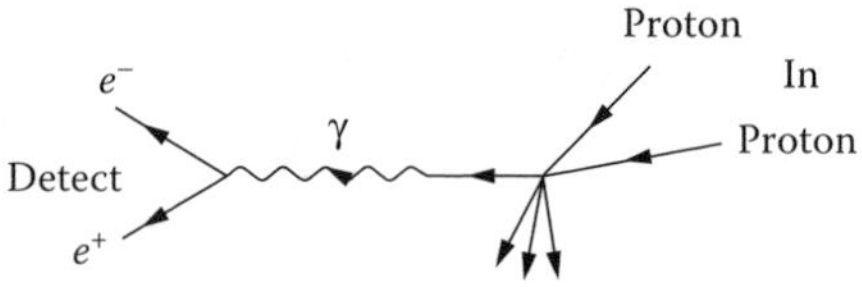

(b)

(b) Ting at Brookhaven produced it by using hadrons and detected its decay into electron and positron.

FIGURE 9.3 J or ψ production: the complementary discoveries of Richter and Ting.

produced an ephemeral meson that subsequently decayed into the electron–positron pair; their total energy in such a case equals the mc^2 of the meson that spawned them.

In a sense, this was Richter's experiment in reverse. Richter sent the electron and positron in at the start and produced the J/ψ meson; Ting produced the meson and detected it by its subsequent decay into an electron and positron. This is summarized in Figure 9.3.

Ting's experiment was a very difficult way to look for mesons. As in all high-energy physics experiments, you have to look at an enormous number of events and perform a statistical analysis on them to see if a peak in the mass histogram is significant or not. You plot the histogram and may find a few extra events in one particular region. As you continue the experiment and accumulate more data, you find this excess consistently appearing. Gradually the small excess builds up into a large amount and then you feel that you might have found something.

Ironically, Ting had been gathering data during the summer of 1974 and found a *huge* enhancement in one small region of the histogram. It was too good to be true. If correct, he had discovered a meson with unexpected stability and substantially heavier than anything ever seen before (its mass of 3095 MeV being more than three times that of a proton). The peak in Ting's distribution was so stunning that at first he suspected that some quirk of the apparatus was artificially producing the enhancement rather than it being a genuine discovery.

During the autumn, Ting's team was checking and rechecking their data. With this intensive work still taking place on the east coast of the United States, Ting sat in a lecture room at Stanford in California and heard that a collaboration of physicists from Berkeley and Stanford had just discovered a meson with precisely the mass and stability that his own data were showing. This confirmed both that he had discovered a genuine effect and that he had to move fast to gain a share of the credit. He named it the J meson and rapidly telephoned his group. The Stanford group named it the ψ meson (it later became known as the J/ψ meson).

The news travelled rapidly around the world and physicists at Frascati in Italy were able to confirm its existence within just a couple of days. Once found, the meson was so visible that it was soon being studied in laboratories the world over.

About 10 days later, the team at Stanford discovered an even more massive particle, the ψ' of mass 3684 MeV. This is formed from the charmed quark and antiquark in an excited state with total spin 1. During the following year, a whole spectroscopy of states built from a charmed quark orbiting around a charmed antiquark was discovered at Stanford and at a similar machine that had just started at Hamburg in Germany. These are called states of 'charmonium.'

Just as hydrogen exists in excited levels, and states of higher energy can decay into states of lower energy by emitting photons, so do the charmonium states of high mass (energy) decay into lower mass charmonium states by emitting photons. The only difference is one of scale: the photon emitted in the atomic case has an energy of a few electron-volts (eV), whereas in the charmonium case, the photon energy is well over a million times larger, being of the order of 100 MeV.

CHARMED MESONS

The spectroscopy of mesons built from a charmed quark and charmed antiquark was uncovered in 1975. The lightest of these (the J/ψ) has a mass of 3095 MeV and so the mass of a charmed quark is therefore of the order of 1500 MeV (Figure 9.4).

The picture was completed with the discovery of 'charmed' particles — mesons built from a charmed quark (or antiquark) accompanied by one of the 'old-fashioned' up, down or strange flavours.

The charmed mesons were discovered in electron–positron annihilation at Stanford about 18 months after the J/ψ. In the aftermath of the annihilation there emerge a charmed meson (e.g., $c\bar{d}$ or $c\bar{u}$) and its corresponding antiparticle ($\bar{c}d$ or $\bar{c}u$). The mass (energy at rest) of the lightest charmed mesons is about 1870 MeV and so to produce such a pair in electron-positron collisions requires a total energy of at least two times 1870 MeV.

The first hint that charmed mesons were being formed was the discovery that when an electron and positron collide with a combined energy greater than about 4 GeV (4000 MeV), the probability of them annihilating increased slightly. This is what would happen if a charmed meson and its anticharm counterpart were being produced. The proof, however, required one of these to be identified specifically. According to the theory that had predicted the existence of charm (p. 124), the charmed quark was a partner of the strange quark in the weak interactions; this implied that when a charmed meson dies, its debris should include a strange meson.

One prediction was that a charmed meson could decay into one strange meson (for example, a K meson) accompanied by a pion. So the experimentalists looked at the debris to see if there was any hint of a K and a π meson being produced, with their energies and momenta correlated in the ways expected if they were the decay products of a pre-existing charmed meson.

Unlike the J/ψ meson, which had been startlingly obvious, charmed mesons were quite elusive and nearly 2 years elapsed before the data were good enough to show

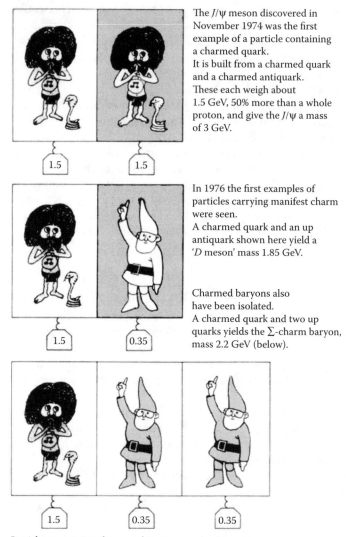

The J/ψ meson discovered in November 1974 was the first example of a particle containing a charmed quark.
It is built from a charmed quark and a charmed antiquark.
These each weigh about 1.5 GeV, 50% more than a whole proton, and give the J/ψ a mass of 3 GeV.

In 1976 the first examples of particles carrying manifest charm were seen.
A charmed quark and an up antiquark shown here yield a 'D meson' mass 1.85 GeV.

Charmed baryons also have been isolated.
A charmed quark and two up quarks yields the Σ-charm baryon, mass 2.2 GeV (below).

Particles containing charm and strange quarks also exist.

FIGURE 9.4 J/ψ and charmed particles: history.

them clearly in 1976. Their masses were about 1870 MeV where the quarks were spinning antiparallel (analogues of the K) and about 2000 MeV when in the parallel spin configuration (analogue of the K^*). Note again the systematic phenomenon where the parallel spin configuration is heavier than the antiparallel; compare p. 102.

So the masses were about right. Furthermore, the charmed meson decayed into strange mesons as predicted. This discovery of charm, and the proof that the charmed quark is the partner of the strange quark, was a major step. It showed that the theories developed over the previous years were indeed on the right track and was an important clue in showing how matter in the universe behaves.

Charmed baryons also exist. The lightest example is Λ_c consisting of cud quarks. Its name recalls the "old-fashioned" Lambda, Λ, made of sud; the subscript c denotes that Λ_c is cud instead of sud. Its mass of 2285 MeV again fits naturally with its constituents: $c(1500)+ud(700)$. As is the case for hadrons made from the u, d, s flavours, the charmed hadrons also have excited states, where the quarks are in S, P, D, etc., levels in atomic physics notation, with correspondingly higher masses.

THE J/ψ SIGNIFICANCE

The discovery of the J/ψ was like finding the proverbial needle in a haystack. Once it had been found, you knew exactly where to look for the predicted related charmed particles and, one by one, they began to emerge with all the properties that had been predicted for them. Thus it was the J/ψ discovery that triggered the breakthrough in unravelling charm.

The charmed particles' existence had been predicted both on aesthetic and also scientific grounds. The aesthetic was that there is a quartet both of leptons and of quarks, and that these formed a common pattern of pairs when partaking in the weak interactions. This gave a tantalising hint that the leptons and quarks are profoundly related to one another (why else should they have such similarity?).

Furthermore, the charmed quark (and consequent particles containing it) had been required in order that ideas uniting the weak and electromagnetic interactions could survive. The discovery that the charmed quark exists *and* with precisely the properties predicted in this theory gave impressive support for the electroweak unification. (Recall that although the neutral currents had been seen in 1973, the discovery of the W and Z carriers of the electroweak force was still, at this stage in 1976, several years in the future. Thus the discovery of charm in 1976 was pivotal in the electroweak story.)

It is therefore no surprise that the leaders of the two teams that discovered the J/ψ (Richter at Stanford and Ting at MIT) were awarded the Nobel Prize for physics within only two years of the discovery.

CHARM: A TESTBED FOR QUANTUM CHROMODYNAMICS

Today it is clear that the discovery of the J/ψ was not only the crucial breakthrough leading to charm and establishing the marriage of the weak and electromagnetic interactions, but also gave impetus to the emerging QCD theory of the strong interaction. One of the most remarkable properties of the J/ψ meson is that it is extremely stable, and it turned out that QCD theory explained this.

As seen already in Chapter 7, the discovery that quarks carry three colours had enabled construction of a theory of their interactions — quantum chromodynamics (QCD).

In 1972, three independent groups of people, Gerhard 't Hooft, David Politzer, and also David Gross and Frank Wilczek, discovered an astonishing property of QCD: the attractive forces clustering quarks into protons and nuclei are not always strong.

At low energies ($\sim$1 GeV) they are powerful, but in *high*-energy particle collisions ($\sim$10–100 GeV) the forces between quarks should be much more feeble. At extreme energies they could become almost free of one another; this property is known as "asymptotic freedom."

All the hadrons known before the discovery of the J/ψ had masses less than 2 GeV, where the up, down, and strange varieties of quarks experienced strong forces. However, the charmed quark was some three to five times heavier than anything previously known, and its discovery revealed a hitherto unknown high-energy world. Today, when experiments producing W and Z with masses approaching 100 GeV are the norm, a charmed particle at 3 GeV seems very low in energy. However, in the 1970s, this was the high-energy frontier and everyone's intuition was based on experiences gained in the "familiar" realm of particles with masses of less than 2 GeV. Thus it was revolutionary both to discover the metastability of the J/ψ and also that QCD naturally explained it: the forces acting on the massive (energetic) charmed quarks are predicted by QCD to be more feeble than the so-called strong forces familiar for the up and down quarks that form protons, neutrons, and nuclei.

A strong attraction between quark and antiquark causes them to meet and annihilate one another. Thus the ϕ meson built from a strange quark and its antiquark soon decays due to their mutual annihilation. The J/ψ meson is an analogous system but built from a massive charmed quark and antiquark rather than the lighter strange quarks. QCD theory predicts that the massive charmed pair is less strongly attracted to one another than is the strange pair. This enfeebled attraction reduces the likelihood of them annihilating, with the consequence that the J/ψ survives much longer than the ϕ. This prediction by Politzer and Tom Appelquist was dramatically verified: the J/ψ lived almost 100 times longer than the ϕ. Thus the elongated life of the J/ψ brought the QCD theory to everyone's attention. This, together with the way that quarks responded to high-energy electron scattering (p. 105) all began to convince physicists that the QCD theory was correct.

Some 32 years later, Gross, Politzer, and Wilczek shared the Nobel Prize for their discovery of the asymptotic freedom of QCD; 't Hooft, who also had stumbled on this, was not included: Nobel Prizes can only be shared by up to three people. In any event, he had already been recognised, along with Martinus Veltman, for establishing the electroweak theory as a correct description of the weak and electromagnetic forces. So in the third quarter of the 20th century, science at last had discovered theories of the electromagnetic, weak and strong forces, which have a common mathematical structure and with which sensible calcuations could be made, and whose implications agree with experiment.

HOW MANY QUARKS?

Our everyday world can be thought of as being built from a hydrogen atom template (an electron and proton system) with the addition of neutrons (forming nuclear isotopes) and the existence of radioactivity where a neutron transmutes into a proton, emitting an electron and also a 'ghostly' neutrino. The electron and neutrino (leptons) appear to be structureless elementary particles. On the other hand, the proton and neutron are now known to be clusters of up and down quarks, which are fundamental spin 1/2 particles along with the leptons.

This pair of up and down quarks has many properties in common with the lepton pair (the electron and the neutrino), and there appears to be a deep connection between them (they have no discernible internal structure, have spin 1/2, respond to electromagnetic and weak interactions in similar ways, etc.). These lepton and quark pairs ($[e^-, \nu_e]$; $[u, d]$) are today known as the 'first generation' of elementary particles.

For a reason that is not yet well understood, Nature repeated itself. The muon (seemingly a heavy version of the electron) also is partnered by a neutrino and shows no sign of any internal structure. Thus there is a 'second generation' of fundamental leptons ($[\mu^-, \nu_\mu]$). Following the discovery of charm, it became clear that there is a second generation of quarks also: the charmed and strange quarks are indeed siblings, connected to one another by the weak interaction so that the charmed quark transmutation into strange is analogous to the down-to-up transmutation that triggered neutron β-decay.

Most physicists had wondered why the muon existed, and whether there were more massive particles akin to the electron and muon. In particular, Martin Perl at SLAC had repeatedly stressed the importance of searching for such entities and it was fitting that in 1975 his team found the first evidence for such a particle. Its mass is almost 2000 MeV (2 GeV), twice the mass of the proton and similar to the masses of the charmed mesons that were also discovered in the debris around that time. Just as the muon is a heavy version of the electron, so does the 'tau' (τ) appear to be a yet heavier version of them both. After 30 years of study, it indeed seems to be a structureless elementary particle, electrically charged and partnered by a neutrino (labelled ν_τ to distinguish it from the ν_e and ν_μ). This $[\tau, \nu_\tau]$ pair seems to be yet another repetition in Nature, and acts like the electron and its neutrino, or again like the muon and its neutrino. Thus we have a 'third generation' of leptons.

As there is now a third generation of leptons, theorists argued that there should also exist a third generation of quarks, to restore the elegant quark-lepton symmetry (Figure 9.5). This new quark pair is labelled t and b for 'top' and 'bottom' (sometimes called 'truth' and 'beauty' though this has fallen from fashion). They were predicted to have electrical charges of 2/3 and −1/3, just as was the case for the first generation (up, down) and the second generation (charm, strange).

As no evidence for particles containing top or bottom quarks had ever been found in low-energy collisions of protons or of electrons and positrons, it was clear that they must be more massive than even charmed particles.

In the summer of 1977, a group of physicists led by Leon Lederman working at Fermilab near Chicago discovered a massive particle known as the Upsilon, Υ, which was produced in proton–proton collisions. Just as the J/ψ meson is a bound state of $c\bar{c}$ (charmed quark and its antiquark) so analogously is the Υ made from $b\bar{b}$. This Υ had a mass of 9.45 GeV, some three times as massive as the J/ψ and ten times the proton!

Lederman's discovery had come in experiments involving collisions between protons and nuclear targets (analogous to the way Ting and collaborators discovered the J/ψ in 1974). In 1977 there was no electron–positron collider with enough energy

The charmed quark completes a second generation of particles. The discovery of the tau lepton and neutrino established a third generation of leptons. The first evidence for a bottom quark came in 1977 and the third generation was completed in 1995 with the discovery of the top quark. Each flavour of quark occurs in three colours. The top quark has been denoted by ? because, although it has been discovered, so little is known about it and its enormous mass raises questions, to this author at least, as to whether it has some unexpected properties. In any event, it is currently an enigma and experiments at the LHC (Chapter 13) will teach us much about it.

1st generation: established 1960s

2nd generation: established 1974–1976

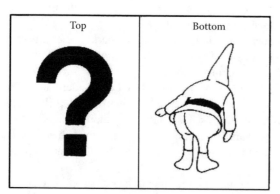

3rd generation: bottom found 1977; top found 1995

FIGURE 9.5 Three generations of leptons and quarks.

to produce an Υ, and it was not until the following year that the Hamburg electron–positron collider, DORIS, was able to confirm it.

The J/ψ had hidden charm, formed from a charmed quark and charmed antiquark binding together to form a meson with no net charm: particles with net charm were subsequently found. Analogously, as the Υ is made from a bottom quark and its antiquark, then mesons built from a bottom quark with an up, down ... antiquark also exist (Figure 9.6), as do 'bottom baryons' such as Λ_b made from *bud*. The first examples of bottom hadrons came in 1980–1981 at CESR (Cornell) and CERN. Today, millions of such bottom mesons are produced at special "*B* factories" as the properties of *B* mesons and their $\bar{B}$ antibottom counterparts contain important information about the asymmetry between matter and antimatter (Chapter 11).

So by 1980 a clear pattern was at last emerging after nearly 30 years of confusion. Electroweak interactions showed that leptons and also quarks formed pairs. Three generations of leptons were known to exist and, but for the absent top quark, there seemed every likelihood that three generations of quarks did also. However, 15 years were to pass before the discovery of the top quark, and the direct proof that it is paired with the bottom quark in the same way that up and down, or charm and strange flavours go together.

The long wait was because the top quark turned out to be immensely massive. It weighs in at some 180 GeV, which is nearly 50 times heavier than the bottom quark, and not far short of the mass of an entire atom of lead. The discovery took place at Fermilab. The experiment involved the collisions of protons and antiprotons at energies of up to 1000 GeV (1 TeV). Under such conditions, the energies in the QCD force fields can cause top quarks and top antiquarks (t and $\bar{t}$) to materialise. They die almost immediately due to β-decay, such as $t \rightarrow be^-\nu$, and a signature for its transient existence was a jet of hadrons coming from the subsequent decay of the *b* quark, accompanied by the high-energy electron from the β-decay process. In addition, there was "missing" energy and momentum, which escaped from the detector, having been carried off by the "invisible" neutrino (see Figure 9.7).

The difference in mass between the 180 GeV of a top quark and the (roughly) 5 GeV of the bottom quark is so huge that a real W^+ is produced when the top quark decays: $t \rightarrow bW^+$. This happens in less than 10^{-24} sec — billions of times faster than in previous experience for processes controlled by the "weak" interaction, such as the β-decay of bottom, charm, or strange quarks where the W existed only ephemerally as a virtual particle. This shows how the feebleness of the "weak" force was actually an illusion caused by the huge mass scale of the W ($\sim$80 GeV) compared to the relatively low masses or energies to which physics had previously been restricted.

With the discovery of the top quark, we have a fundamental particle that is more massive than either the W or Z. Its lifetime of less than 10^{-24} sec is so short that a t or $\bar{t}$ is expected to die before having a chance to bind to one another. So there is unlikely to be a $t\bar{t}$ analogue of the $J/\psi(c\bar{c})$ or $\Upsilon(b\bar{b})$ unless some hitherto unknown force or phenomenon takes over at such extreme energies.

There is no electron–positron collider capable of producing energies where top quarks and antiquarks could appear, though there is intense international discussion about such a machine being built in the future. The Large Hadron Collider (LHC) at CERN collides protons head-on at energies up to 10 TeV, and will produce large

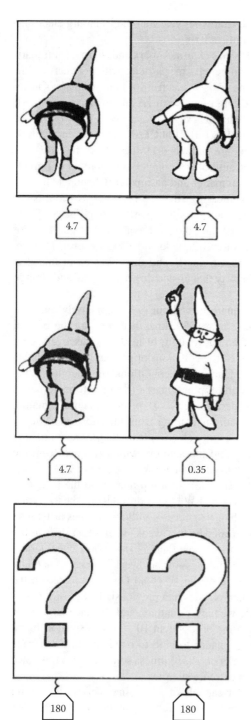

The Υ is built from a bottom quark and bottom antiquark analogous to the way the $J\,\psi$ is made from charm and anticharm. The 9.4-GeV mass arises from the two bottom flavours, each being about 4.7 GeV.

"Bottom" mesons with masses just over 5 GeV exist, built from a bottom quark (or antiquark) with either up or down antiquarks (or quarks).

Will a meson made of a top quark and top antiquark exist? It would have a mass of around 360 GeV if it did. However, theory suggests that its constituents decay into lighter flavours before having been able to grip one another to form the meson. That is another example of the enigmas related to top quarks and why we put question marks here.

FIGURE 9.6 Bottom and top quarks.

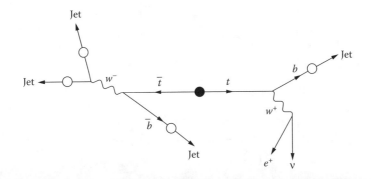

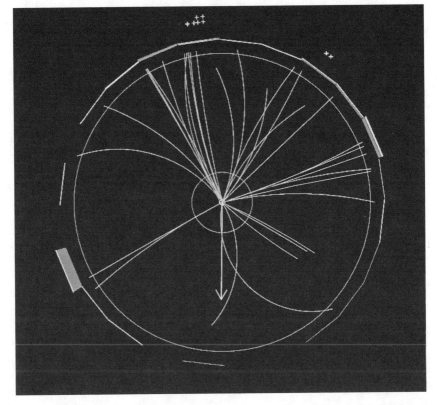

FIGURE 9.7 Top Discovery. A proton and antiproton have collided to produce a top quark and antiquark in the CDF detector at Fermilab. They each experience weak decay: $t \rightarrow bW^+$ and $\bar{t} \rightarrow \bar{b}W^-$. The b and $\bar{b}$ each produce a jet of hadrons; the $W^- \rightarrow u\bar{d}$ gives two further jets, making four in all. The $W^+ \rightarrow e^+\nu$ gives a positron, which is recorded when it annihilates and deposits energy in the outer region of the detector (leaving the gray block at 8 o'clock), and the direction of the neutrino is shown by the arrow. Adding the energies and momenta of these jets and particles gives the mass of the original top and antitop particles.

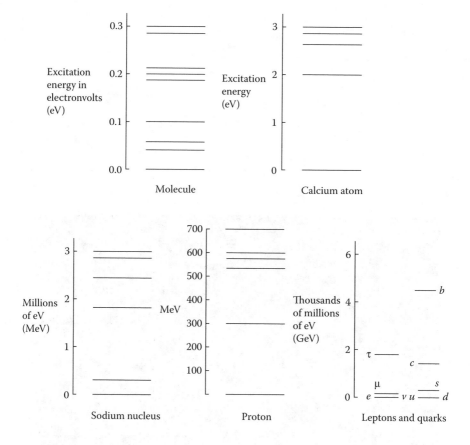

FIGURE 9.8 Energy levels for excited states of molecules, atoms, a nucleus, and a proton. Note how the smaller the structure, the greater the scale of excitation energy. The excitations correspond to rearrangements of the constituents. The top quark is so massive that its position would be another four page heights above the top of this page. Why the quarks and leptons have such a pattern of masses is presently unknown.

numbers of top quarks. One of the exciting questions that the LHC experiments may answer is whether there are unexpected properties of the massive top quark or whether it is "simply" the final link in the pattern of fundamental fermions — leptons and quarks.

Thus we now know that Nature contains a third generation of quarks partnering the third generation of leptons [τ, ν_τ]. As things stand at the start of the LHC experiments, we have found three complete generations of fundamental particles of matter (why we believe that it is only three will be described in the next section). This is an essential part of what is now known as the Standard Model. These discoveries confirm that we are on the right track with our embryonic theories attempting to unify the weak, electromagnetic, and strong interactions, and also for intensifying the suspicion that quarks and leptons are intimately related to each other.

In a very real sense, the recognition of the generation patterns for quarks and leptons is like the discovery of the Eightfold Way (Chapter 5) for hadrons and Mendeleev's periodic table for atoms. The discovery of predicted top quarks was analogous to that of the Ω^- discovery in the Eightfold Way and to the discovery of the elements gallium and germanium in the periodic table. Having confirmed the validity of the pattern, the question today is: 'What causes the pattern to exist?' In both the periodic table of the elements and the Eightfold Way for hadrons, the cause was a deeper layer in the Cosmic Onion. In the case of quarks and leptons, however, no hint has emerged of an analogous substructure: they appear to be "pointlike" down to distances as little as 10^{-19} m — some 10,000 times smaller than a proton — which is the limit of resolution that experiments can currently reach. It is suspected that their relationship and the cause of the generation patterns is more profound. In particular, each generation appears to be identical to the others in their electrical charges, their response to the weak and strong interactions, their spins, and other features except for one: their masses are very different. A single top quark is more massive than an entire atom of gold, while an up quark is lighter than a proton; a τ weighs as much as an atom of heavy hydrogen (deuterium), while an electron is some 2000 times lighter than hydrogen, and its neutrino is so light that a precise value for its mass is still awaited. Understanding the origin of mass, through the Higgs mechanism, is thus one of the major challenges at the start of the 21st century, as is finding the explanation for their remarkably varied values. (See also Figure 9.8.)

10 The LEP Era

By 1989 the Electroweak and Quantum chromodynamic theories of the forces, and the generations of leptons and quarks (apart from the top quark whose discovery was still 5 years away), gave the essence of what has become known as the Standard Model. This was a great advance compared to where the frontier of physics had been only 15 years earlier, but, in turn, raised new questions such as why do the particles have their special masses? Why are the strengths of the forces as they are? And what differentiates the generations other than mass? There were also questions about the menu of fundamental particles, not least: Does top really exist? Are there further generations? And are quarks and leptons fundamental or is the pattern of generations a hint that there are yet more basic constituents so that matter is constructed like a set of Russian dolls?

These were what concerned physicists as LEP — the Large Electron Positron Collider at CERN — began operation in July 1989. This was a ring of magnets deep underground, 27 km in length, which accelerated electrons in one direction and positrons in the other. At four points around the ring, the two counter-rotating beams were brought into head-on collision so that electron and positron could mutually annihilate.

From 1989 until 1996 the beams in LEP had energies of around 45 GeV each, chosen so that their annihilation took place at a total centred on 90 GeV, the energy at which the Z^0 could be produced: LEP had been designed as a Z-factory. Over the next 10 years it made Z particles some 10 million times, enabling the Z to become one of the most intensively studied particles of all. If the Z^0 (and by implication the $W^{\pm}$) was other than simply a heavy version of the photon (e.g., did it have an internal structure?), such scrutiny would show it.

It would also be possible to find out how Z^0 decayed. The Standard Model predicted the lifetime of Z^0 and the nature of what is produced in the debris of its demise, so testing these predictions to the best precision might reveal chinks in the theory. Also, as mirror symmetry (parity) was known to fail in β-decays and in processes involving the W particle, LEP could investigate how Z^0 behaved under mirror symmetry; here again, the Standard Model specified what was expected.

It was possible to vary the energy of the beams so that collisions occurred over a range of energies around 90 GeV. The chance of them annihilating changed as the energy varied. When the energy was just below 90 GeV, the resulting "cross section" was found to be small; it then rose to a peak at around 91 GeV before falling away again at higher energies. This bump is due to the formation of Z^0 in the e^-e^+ annihilation. Before the advent of LEP, the mass of Z was known to an accuracy of 2%, with value 92 ± 2 GeV, i.e., could be as high as 94 or as low as 90 GeV; as a result of the measurements at LEP, where the position of the peak could be determined very accurately, the mass is now known with a precision of better than one part in ten thousand. Its value is

$$m(Z^0) = 91.188 \pm 0.002 \ \text{GeV}$$

The bump in the cross section is centred on this particular value, and spread around it, rising to the peak and then falling away again over a range of a few GeV energy. The amount of this spreading, known as the width, shows how long Z^0 lives before decaying into more stable particles. This is how.

The uncertainty principle (Table 2.4) limits the accuracy with which one can simultaneously know both the momentum and position of a particle. It also limits our ability to know both the energy and time of an event such that the product of the two uncertainties — Δt for the time (in seconds) and ΔE for the energy (in GeV) — is

$$\Delta t \times \Delta E \geq \hbar = 6.6 \times 10^{-25} \text{ GeV-sec}$$

If the Z were absolutely stable, you would be able to measure its properties for as long as you wish. Suppose you took an infinite amount of time, $\Delta t \to \infty$; the above equation would then imply that its mass, or intrinsic energy mc^2, could be measured to perfect precision, $\Delta E \to 0$. In such a case the bump would have been a thin spike at the point where the beam energies coincided with its rest mass. In reality, the Z^0 has a finite lifetime, which limits the Δt available; the above equation then implies an uncertainty in E that is at least $\hbar/\Delta t$. Hence, in place of a thin spike at a precise value of E in the cross section, a Z^0 particle with a finite lifetime will be revealed by a spread in this energy. The uncertainty in energy, ΔE, is measured by the width of the bump at half of its height, which LEP showed to be 2.5 GeV. Putting this into the above equation shows that $\Delta t \sim 10^{-25}$ sec.

Having determined the lifetime of the Z^0, the challenge was to compare it with what theory predicted. This is where the first dramatic discovery was made.[1]

The Standard Model predicted that the Z decays democratically into each lepton and its antilepton, and into each colour and flavour of quark and its corresponding antiquark. Thus the three colour possibilities available for each flavour of quark and antiquark — $q_R\bar{q}_R$, $q_B\bar{q}_B$, $q_Y\bar{q}_Y$ — makes decays into $q\bar{q}$ some three times more likely than into the colourless leptons. This was verified; the propensity for the Z to decay into jets of hadrons, triggered by a specific flavour of quark, $e^+e^- \to Z^0 \to q\bar{q}$, is some three times more likely than for it to decay into a pair of leptons such as $e^+e^- \to Z^0 \to \mu^+\mu^-$.

By detecting the flavours of the emerging hadrons, where short-lived examples containing b flavours decayed visibly in the detector, or strange mesons such as the K^0 decay in a characteristic V shape, the democratic affinity for the quark flavours was established. Its affinity for leptons was easier to test as $Z^0 \to e^+e^-; \mu^+\mu^-; \tau^+\tau^-$ were directly visible in the detectors. However, neutrinos left no trail; $e^+e^- \to Z^0 \to \nu\bar{\nu}$ is in effect invisible. If you could have been sure that an annihilation had occurred, then you could have inferred the $e^+e^- \to Z^0 \to \nu\bar{\nu}$ by the apparent disappearance of energy; however, there are billions of electrons and positrons in the beams, most of which miss each other and only occasionally do two meet and mutually annihilate, so there is no direct way to know that $e^+e^- \to Z^0 \to \nu\bar{\nu}$ has happened. Nonetheless,

---

[1] At SLAC, micron-sized beams of positrons and electrons created in their linear accelerator were brought into collision at around 90 GeV, enabling them also to measure the lifetime of the Z. The SLC (Stanford Linear Collider) eventually made about 100,000 Z while LEP accumulated 10 million.

it is possible to know, on the average, that this has taken place and to measure how often. This has to do with the lifetime of the Z^0.

According to the Standard Model, the Z^0 can decay democratically into all varieties of neutrino. Thus $Z^0 \rightarrow \nu_e \bar{\nu}_e$; $Z^0 \rightarrow \nu_\mu \bar{\nu}_\mu$; and $Z^0 \rightarrow \nu_\tau \bar{\nu}_\tau$, each of which occurs with the same probability and at a rate relative to that of their charged lepton counterparts, $e^+ e^-$, $\mu^+ \mu^-$, $\tau^+ \tau^-$, which is specified by the theory. So, first the experimentalists measured and verified the rates for the latter set; knowing this, they next calculated how probable decays to neutrinos should be and then, adding all of these different decay possibilities together, computed how long the Z^0 should live and hence what its width should be. Finally one can compare this theoretical number with what the experiment measured. The result was remarkable.

The theoretical lifetime will depend on how many varieties of light neutrino there are. Suppose that there are further generations of quarks and leptons akin to those we already know but so massive that they have, to date, escaped detection. If they are like the known three, in that they each have a light, almost massless, neutrino, then the Z^0 will decay into each of these varieties, $Z^0 \rightarrow \nu_i \bar{\nu}_i$ where the label i will refer to each and every generation that exists. So when $i = 1, 2, 3$, we have the neutrino partnering the electron, muon, and tau; when $i = 4, 5$, etc.; we have the neutrinos partnering leptons as yet undiscovered (and if the pattern is a guide, of quarks, too). The more varieties there are, the faster the Z^0 will decay, the shorter is its lifetime and the broader its peak. Figure 10.1 shows how the Z^0 appears as a peak in the rate that electron and positron annihilate as their energy varies. As their total energy increases from 88 to 95 GeV, the rate rises and falls (black dots in the figure) with a peak when the energy (bottom axis) is just above 91 GeV. The curves correspond to what the shape would be if there are two (upper curve), three (middle curve), or four (lower curve) varieties of light neutrino. After these measurements of more than 10 million Z, the match between the shape of the bump in the data and the middle curve, which was predicted for three varieties of neutrino, is indisputable.

Thus the Z^0 has revealed that there are only three varieties of light neutrino, at least of the variety that the Standard Model recognises. This does not rule out the possibility that there are not extremely massive neutrinos, or "sterile" neutrinos that behave differently than those with which we have become familiar, but it is nonetheless a profound result that there only three generations of fundamental leptons and quarks of the form that the Standard Model recognises. This is a strong hint that these particles are not made of more basic entities (for if they were excited forms of a deeper layer of the Cosmic Onion, we would expect the pattern to continue onwards and upwards).

The nature of neutrinos, and the question of whether they have any mass and if so how much, is one of the main questions that has come from the LEP era (see Chapter 12 for more about neutrinos).

QUANTUM FORESIGHT

The Z^0 decays democratically into any variety of particle and its corresponding antiparticle so long as their combined mass is less than that of the Z: energy conservation prevents heavier particles from being produced. So the Z can decay into all varieties of lepton and into the following flavours of quarks: $u\bar{u}$, $d\bar{d}$, $s\bar{s}$, $c\bar{c}$, and $b\bar{b}$. It cannot

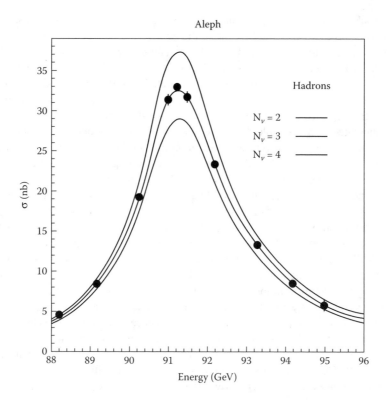

FIGURE 10.1   The Z peak and three neutrinos. The data are the black dots. The three curves are predictions for the shape of the Z^0 production rate if there are two, three, or four varieties of light neutrino. The middle curve, corresponding to three neutrinos, fits the data perfectly. (*Source*: ALEPH collaboration/CERN.)

decay into the top quark and its antiquark, $t\bar{t}$, as their combined mass of some 360 GeV far exceeds that of the Z^0. However, due to the quantum uncertainty phenomenon, the top quark can nonetheless affect the behaviour of $e^+e^- \to Z^0$ in measurable ways.

While energy conservation is only true over long time scales, the quantum uncertainty relation

$$\Delta E \times \Delta t \geq \hbar$$

implies that for brief times of duration Δt, the energy account can be overdrawn by an amount ΔE. So, imagine running LEP at 91 GeV, the energy where the Z^0 is produced at peak rate (Figure 10.1). The extra energy that would be needed to produce a top quark and its antiquark, $t\bar{t}$, would be about 270 GeV. The above relation implies that you can "borrow" this amount for a time $\Delta t \sim 10^{-27}$ sec. This is a fraction of the time that the Z^0 itself lives, so there is the possibility that during its short lifespan it could transform momentarily into a $t\bar{t}$ — for no longer than a brief 10^{-27} sec — and then back to a Z again. These so-called "quantum fluctuations" will leave their imprint on the ephemeral Z and be manifested in some of the experimental measurements (see Figure 10.2).

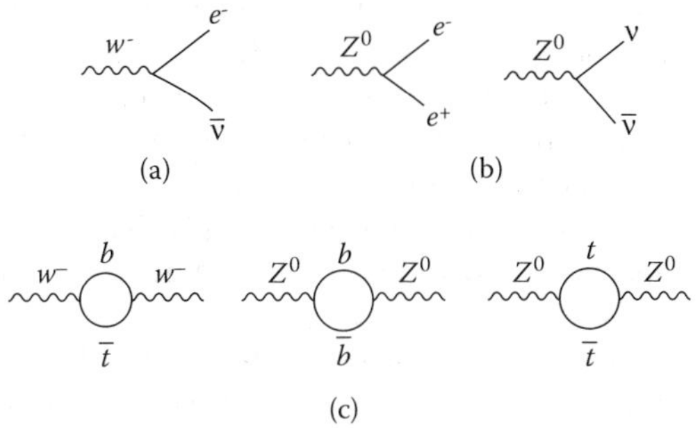

FIGURE 10.2 Quantum foresight: Top quarks affecting the W and Z. In (a) the W^- decays to $e^-\bar{\nu}$ in the familiar β-decay manifestation of the weak force. In (b) the neutral manifestation of the force sees the Z^0 decay to e^-e^+ or $\nu\bar{\nu}$. The relative strength of these two is predicted by the Standard Model to be the same when factors related to the difference masses of W and Z, which in turn relate to θ_W (Figure 8.3), are taken into account. The early measurements of these relative strengths agreed with them being the same but as data from LEP accumulated and, with this, the precision improved, a subtle deviation began to emerge. This difference is due to "quantum bubbles" (c) $W \to b\bar{t}$ and $Z^0 \to b\bar{b}$ and $t\bar{t}$. The large mass of the top is felt twice in the $t\bar{t}$ but only once in the $W \to b\bar{t}$. This causes a subtle difference in the quantum effects for the W and Z. If the mass of the top quark were infinite, the effect on $W \to b\bar{t}$ would vanish, as would that for $Z \to t\bar{t}$; however, that of the $Z \to b\bar{b}$ would remain. So we can guess that the the effect of the top mass, m_t, differs for the W and Z. Precision measurement of the strength of the weak force as manifested by the Z^0 was one example of how the top quark was "felt early" even though there was not enough energy in the LEP experiments to produce it directly.

The Standard Model determines the couplings between the Z and the transient $t\bar{t}$, which enables the implications of the quantum fluctuations to be computed. Their impact on the properties of the Z^0 depends critically on the mass of the top quark. When LEP began in 1989, no one knew for sure that the top quark existed, let alone how massive it might be; but as information on the Z accumulated, the precision improved to a point where it became sensitive to the minute effects of the $t\bar{t}$ fluctuations. By 1994 the measurements from LEP had become so accurate that theorists could deduce that top quarks probably existed, and with a mass that was somewhere in the range of 150 to 200 GeV, most probably around 170 to 180 GeV.

Although production of such massive particles directly was out of LEP's energy reach, experiments with beams of protons and antiprotons at Fermilab (near Chicago) whose energies reached up to 1 TeV (1000 GeV), could make them. The LEP results were brilliantly confirmed in 1995 when Fermilab produced the top quark, and showed its mass to be 179 ± 5 GeV. The collisions produced a handful of top quarks, each of which was detected by β-decay into the bottom flavour (Figure 9.7), as predicted by the Standard Model.

From 1997, LEP operated at its maximum energy of just over 100 GeV per electron and positron, a total of 200 GeV in their annihilation. This is still far short of what is required to create top quarks ($e^+e^- \to t\bar{t}$ would need at least 360 GeV in total) but is enough to produce the charged siblings of the Z^0: the $W^\pm$.

The conservation of electric charge enables a single Z to be produced in $e^+e^- \rightarrow Z^0$, and hence a total energy of 91 GeV is sufficient for that task. However, the electrically charged W bosons must be produced in matching pairs, $e^+e^- \rightarrow W^+W^-$, in order to preserve the overall electric charge of the process. This requires enough energy to produce *two* W, totalling some 161 GeV or more.

The conservation of energy and momentum imply that if the e^+e^- annihilation takes place below this energy, no W pairs will emerge. At an energy of 161 GeV, it is just possible to produce a pair where each of the W is at rest. As the collision energy of the electron and positron is increased, the total will exceed that required to make two W at rest; the spare energy may lead to the creation of further particles, such as photons, or be transformed into kinetic energy of the two W. The probability of this happening increases rapidly as the total energy of the collision increases. By mapping out how the production rate varies with energy, it is possible to make an accurate measurement of the W mass.

The behaviour turned out to agree exactly with the prediction of the Standard Model, where the mass of the W is 80.42 GeV, with an uncertainty of only 0.04 GeV. Fermilab has also produced large numbers of W in collisions of protons and antiprotons (the same types of experiment with which they produced the top quarks). These have shown that the lifetime of the W is some 10^{-25} sec, similar to that of the Z, as the Standard Model predicted.

Everything so far about the Z and W appears as expected: they show every sign of being the fundamental carriers of the weak forces, and partners of the massless photon in the Standard Model. Armed with this assurance and accurate values for their masses, as well as the mass of the top quark, it is possible to revisit the quantum foresight calculations of the Z and W properties. It turns out that the masses and lifetimes of the Z and W do not quite correlate well with one another if, and this is the big "if," quantum fluctuations involving only the top quark (or antiquark) are at work. To describe the totality of the data, it appears that some additional quantum fluctuation is taking place: some other particle, as yet undiscovered, appears to be heralding its existence by its effects on the Z and W. The questions are: What is it, and how might we produce it in experiments?

Far from heralding some failure in our world view, this subtle mismatch is widely suspected to be the first manifestation of the Higgs boson, which is the generally accepted missing piece in the Standard Model. The successful predictions of the existence and masses of the W and Z, and the formulation of the electroweak theory itself, relied upon the "Higgs mechanism." This theory, due to Peter Higgs, implies that there is a spinless particle, known naturally enough as the Higgs boson, and that it is the strength with which this interacts with various particles that determines their masses. In a chicken-and-egg sort of fashion, it also generates a mass for itself. The theory cannot say precisely how big this is, only that it is less than about 1 TeV (1000 GeV); for more on this, see Chapter 13.

If it is this heavy, then to be able to produce the Higgs boson in experiments will require the energies of collisions at the Large Hadron Collider (the 7-TeV proton collider in the tunnel that originally housed LEP and starting operation in 2007). However, the Higgs boson could be much lighter than this, perhaps even 100 to 200 GeV, and, as such, be on the borderline of appearing before the LHC is fully

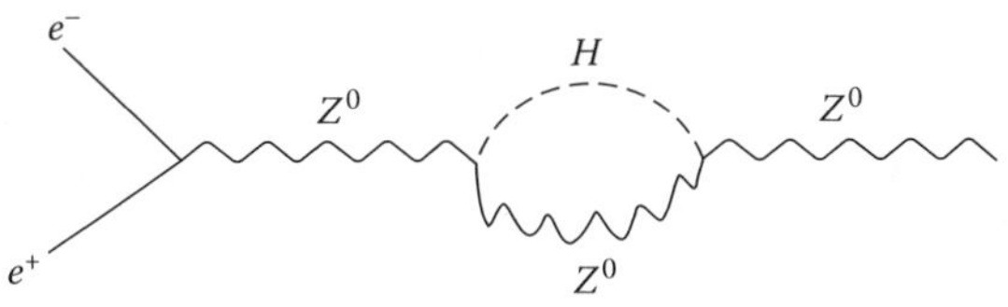

FIGURE 10.3 Quantum foresight: the Higgs. Electron and positron annihilation at a total energy of 91 GeV produces a Z^0. A Higgs boson, denoted H^0, momentarily appears and disappears. This can affect the measured properties of the Z^0 in this reaction.

operational or even of showing up, like the top quark, courtesy of quantum foresight. Could it be that the subtle discrepancies in the Z and W properties are the first hint of the Higgs boson, effervescing in quantum fluctuations?

Theory would suggest that this is so. The Higgs boson, as the bringer of mass, has greater affinity for coupling with heavy particles than with light ones. With the exception of the top quark and the Higgs boson itself, the Z is the heaviest known and as such the one for which the Higgs boson has the strongest coupling. This causes the following fluctuation to be possible (Figure 10.3):

$$e^+ e^- \rightarrow Z^0 \rightarrow Z^0 + H^0 \rightarrow Z^0$$

The effects predicted are in accord with what is seen. However, unlike the case of the top quark where the magnitude of the phenomenon was sensitive to the mass of the top and an accurate (and successful!) prediction of its mass could be inferred, in the case of the Higgs boson, the dependence on its mass is rather slight. So whereas for the top it was possible to give a precise pointer in advance of its eventual discovery, for the Higgs boson all we can say from the above measurement is that there is a Higgs boson at 200 GeV or below. (It is even possible that there might be a *family* of Higgs boson particles, and that this is but the lightest — time and experiment will tell).

The fact that we can already put an upper limit as "low" as 200 GeV is very exciting. When one examines the theory and the data more carefully, one finds that within the range of possibilities, the region 110–130 GeV is most likely. This is tantalisingly near the maximum that LEP was able to reach, and in its final months every extra bit of energy was put into the experiments in the hope of producing a Higgs boson. In this they were unlucky; the Higgs boson (as of October 2006) is still awaited. It might be produced at Fermilab (if it is in the lower part of the 100–200 GeV range), or at the LHC after 2007 if it is heavier than this.

11 *CP* Violation and *B*-Factories

The experiments at LEP showed what was happening in the universe when it was less than a nanosecond old. They revealed a state of symmetry where the energy from the still-hot Big Bang is congealing into particles of matter and antimatter in perfect balance. In all experiments performed so far, under these conditions there is no preference for one form (such as matter) over the other. This is utterly different from what we see today, billions of years after the Big Bang, where the universe is dominated by matter and radiation to the exclusion of antimatter. The cause of this asymmetric situation is one of the greatest unsolved mysteries in science.

An "easy" solution would be to assert that there was an excess of matter to begin with. However, this begs the question of "Who ordered that?" and avoids the problem rather than solving it. Furthermore, there are hints that particles of matter and antimatter do not always behave symmetrically. This has stimulated ideas on how unstable particles and their antiparticles may die asymmetrically such that a dominance of matter might emerge.

In 1966, Andrei Sakharov realised that in a cooling and expanding universe, two essential conditions are needed for an imbalance between matter and antimatter to emerge spontaneously. First, protons must decay, but so slowly that, in the entire history of the Earth, the totality of such decays would amount to no more than a few specks. The second condition is that there must be a measurable difference between matter and antimatter.

Regarding the first of these, most theorists suspect that protons are unstable, though experiment has so far been unable to detect this most exceptional phenomenon. The profound similarities between quarks and leptons that we have met inspire belief in some grand unified theory. Even though there is no concensus on the precise details of such a theory, a general consequence of uniting leptons and quarks seems to be that protons, being made of quarks, can decay into a positively charged lepton, such as a positron, and emit energy in the form of photons. While Sakharov's first condition remains theoretical conjecture, it is his second condition where exciting progress has taken place recently and which is the main theme of this chapter.

Sakharov was inspired to his insight following the discovery of a subtle difference between certain strange particles, the kaons, and their antiparticle counterparts. Technically this showed that a particular symmetry, known as *CP invariance*, is not a general property of Nature. This provided the first and, until recently the only, example of a manifest asymmetry between particle and antiparticle. In order to understand the implications, we will need to describe what *CP* symmetry is.

CP SYMMETRY

Replacing a particle by its antiparticle involves changing attributes such as electric charge, strangeness, and other flavours from positive to negative (or vice versa). This action is known as charge conjugation, denoted by C. Viewing something in a mirror, which is equivalent to changing all directions in space, front to back, left to right, is known as parity, denoted by P. Doing each of these together is known as CP.

After the discovery that parity symmetry does not occur when the weak force is at work, the belief developed that CP might be the true symmetry. For example, neutrinos always spin in the same direction, which is conventionally called left-handed, and their mirror image would be a right-handed neutrino. As right-handed neutrinos do not exist in the real world, the production of neutrinos does not exhibit parity symmetry. However, if we now imagine the effect of charge-conjugation, C, this will change that (right-handed) neutrino into a (right-handed) antineutrino, which does exist in the real world. Thus, although the behaviour of neutrinos and antineutrinos satisfies neither parity or charge conjugation separately, the combined action of CP symmetrically matches neutrinos with antineutrinos. The cultural shock that parity and charge-conjugation symmetries fail in processes controlled by the weak interaction was replaced by a new paradigm: for any process that happens in Nature, change all the particles to antiparticles and view it in a mirror, and what you see will be indistinguishable from the real world. However, this belief was dashed in 1964 by the discovery that CP symmetry is not always exact. An implication is that particles and antiparticles, matter and antimatter, have essential differences.

The discovery of CP *violation* involved the strange K mesons, and until recently these were the only known examples of the mysterious phenomenon. The ingredients of our story, which leads to modern ideas about the asymmetry between matter and antimatter, will be these.

CP violation was a complete enigma when discovered in 1964 and would remain so for many years. Theorists came up with ideas, most of which led nowhere, among which was one in 1973 by two Japanese physicists, T. Kobayashi and M. Maskawa, who have become famously abbreviated to "KM." Their idea initially made little impact as it seemed to bear no relation to reality, at least as it was then perceived. One way of summarising their idea is to say that they made a host of assumptions for which there was no evidence, in order to explain a single phenomenon. Theoretical papers that do this most often end up in readers' wastebaskets, and this seems to have been largely true in 1973. However, the chain of experimental discoveries that began with charm in 1974 changed all that, as one by one KM's assumptions turned out to be verified. So what is it that they had done?

With modern hindsight we would say that the reason CP violation occurs for the K mesons is because they contain constituents from different generations. Such a concept would have been impossible to articulate when KM wrote their paper in 1973, a year before the discovery of the charmed quark, as no one at that time even realised that "generations" are an integral part of the particle pattern. Only three flavours were then known, though the fourth had been hypothesised. What KM suggested was that there be a third pair of quarks in addition to the (hypothesised) two pairs of down/up and strange/charm. Building on work by the Italian theorist Nicola Cabibbo, they realised that they could summarise the responses of the different types of quark to the weak interaction by numbers expressed in a 3×3 matrix. This is now known as the CKM matrix (after the three theorists). If certain of these numbers are complex (involve the square root of -1), *CP* violation can arise.

Their work would have remained a curiosity but for the discovery of charm in 1974, a fifth quark (bottom) in 1977, and the sixth quark (top) in 1995 giving the third pair or "generation." The discovery that there are indeed three generations galvanised interest in the CKM theory. The question that grew in the minds of many theorists was whether the existence of three generations of quarks is what has led to the dominance of matter in our universe. The theory also implied that *CP* violation, while only a trifling and subtle effect for the strange mesons, should be large for *B* mesons, where the strange quark or antiquark of the *K* meson is replaced by a bottom flavour. As the 21st century began, the evidence began to suggest that they are correct: *B* mesons do exhibit large violations of *CP* symmetry. At experiments at the LHC from 2007, billions of *B* mesons will be produced. One dedicated detector, known as LHCb, has been designed with the task of studying this phenomenon.

The remainder of this chapter describes these ideas in more detail. Readers who are not interested in this can move on to Chapter 12 without missing any essential parts of the story.

THE STRANGE CASE OF THE NEUTRAL KAONS

The main character in the *CP* symmetry story is the electrically neutral strange particle — the kaon — made from a quark and antiquark. Although kaons are not simply matter or antimatter, in that they contain both a quark and an antiquark, their inner structure is unbalanced between matter and antimatter and it is that which makes them of special interest.

I say "unbalanced" because the quark and antiquark within the kaon have different flavours and come from different generations (p. 134) — a d (or $\bar{d}$) from the first generation and $\bar{s}$ (or s) from the second. For example, the electrically neutral kaon, K^0, is built from a down quark joined to a strange antiquark; hence, $K^0 \equiv d\bar{s}$. This combination can be thought of as "matter biased." The antikaon is the exact reverse of this. It has a down antiquark cohabiting with a strange quark, $\bar{K}^0 \equiv s\bar{d}$. If both the *C* and *P* operations take place, a kaon will be turned into an antikaon and vice versa, as Figure 11.1 illustrates.

In order to get back to where you started, you would have to do the *CP* operation twice: doing it once turns $K \rightarrow \bar{K}$ and the second turns $\bar{K}$ back into K. Mathematicians would denote this as $(CP)^2$, and the action of starting with a kaon and ending with a kaon is equivalent to having multiplied the mathematical expression describing the original kaon by 1. If $(CP)^2$ is equivalent to the identity, 1, then the effect of acting with *CP* once can be represented by either $+1$ or -1, as either of these times itself leads to unity. If instead of the K^0 we had a π^0, where the flavour of quark and antiquark are matched, $u\bar{u}$ or $d\bar{d}$, performing the *CP* operation just once will be sufficient to get you back to where you started. In the language of quantum theory, the π^0 is said to be an "eigenstate" of *CP*. The ways in which pions are produced

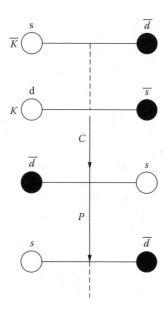

FIGURE 11.1 CP for π and K. CP symmetry is easiest to illustrate for a quark and antiquark with the same flavour. So let us choose $u\bar{u}$ as in an electrically neutral pion π^0, and define the u to be on the left and the $\bar{u}$ on the right side of the midpoint. C then takes $u\bar{u} \to \bar{u}u$; in the mirror, P will in effect swap their positions $\bar{u}u \to u\bar{u}$ and we are back where we started. So the CP operation has taken the π^0 into itself. We say that it is an "eigenstate" of CP. For a kaon (or antikaon), the CP operation would take you to an antikaon (kaon), respectively. This is illustrated in the figure, where the effect of C (quark turned into its "negative" image antiquark) and P (mirror reversal) is shown. The kaon has in effect been turned into an antikaon.

in strong interactions, for example, in the decays of heavier mesons such as the rho or omega, show that for the π^0 the CP value is -1. (This number is known as the eigenvalue.) For a collection of pions that has no electrical charge overall, the value of CP is the product of the individual "minus-ones;" so for an even number of pions this is $+1$, and for an odd number of pions it is -1. This has been tested and repeatedly confirmed in more than half a century of experiments.

Suppose instead of a kaon or an antikaon we had something that is a 50:50 mixture of each. This may sound bizarre but in quantum mechanics such a split personality is quite normal. We could call the two states

$$K_S^0 \equiv K^0 + \bar{K}^0; K_L^0 \equiv K^0 - \bar{K}^0$$

(The subscripts S and L refer to "short" and "long," for reasons that will become clear later).

Now, as $CP \times K^0 \to \bar{K}^0$ and $CP \times \bar{K}^0 \to K^0$, see what happens when CP acts on either of $K_{S,L}$:

$$CP \times K_S^0 \to \bar{K}^0 + K^0 \equiv K_S^0$$

whereas

$$CP \times K_L^0 \to \bar{K}^0 - K^0 \equiv -K_L^0$$

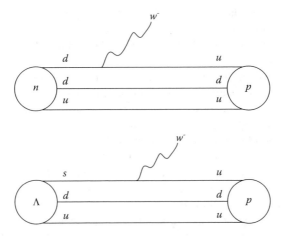

FIGURE 11.2 Beta decay which destroys strangeness. The β decay of a neutron, caused by $d \to uW^-$, has an analogue with the strange flavour $s \to uW^-$, which destroys strangeness. This is illustrated for $\wedge$ (sdu); an analogous process causes K° decay.

So we can summarise these as follows:

$$CP \times K_S^0 = +K_S^0; CP \times K_L^0 = -K_L^0$$

which shows that after the CP operation on either of these states, we get back that state, but with a $+1$ or -1 overall sign in front. Hence, the K_S state is revealed to have CP eigenvalue of $+1$ and the K_L state has eigenvalue -1 (see Figure 11.1).

The kaon is the lightest meson containing a strange quark or antiquark, but when it decays, the strangenesss disappears. The simplest way to see how is to look at the electrically neutral $K^0(\bar{s}d)$ or its antiparticle, $\bar{K}^0(s\bar{d})$. This can undergo a form of beta decay where $\bar{K}^0(s\bar{d}) \to \pi^+(u\bar{d})e^-v$. What happened is that at quark level $s \to u(+e^-+v)$ analogous to the traditional beta-decay triggered by $d \to u(+e^-+v)$ (see Figure 11.2).

When a $\bar{K}^0(s\bar{d})$ decays, its $s \to u$ by the weak interaction discussed earlier; the electric charge is preserved by the emission also of $e^- + \bar{v}$ or alternatively by the appearance of $d\bar{u}$. So what begins as $s\bar{d}$ can end up as $u\bar{u}d\bar{d}$, which are combinations of quarks and antiquarks that we find within pions. So a K^0 can decay to two or more pions. This is indeed what happens in practice and is where CP symmetry now enters the story.

If CP is a symmetry in Nature, the value of $+1$ or -1 will be preserved throughout the decay, before and after. The K_S^0 variety has CP eigenvalue of $+1$ and is thus allowed to produce two (an even number) pions.

What about the K_L^0? This has CP eigenvalue of -1 and so to preserve this CP eigenvalue, must convert into an odd number of pions when it decays. A kaon cannot transmute into just a single pion as energy and momentum cannot be conserved, so *three* is the minimum odd number that K_L^0 can produce. Having to make three pions is more difficult than simply making two, with the result that the K_L^0 resists decay longer than does K_S^0. This is indeed what is found: the electrically neutral kaon either dies in about one-tenth of a thousandth of a millionth of a second, or has an elongated life of some five hundredths of a millionth of a second. These are known as the "short-" and "longlived" modes, respectively, and hence the traditional labels K_S^0, K_L^0 as we have used.

In 1964, James Christenson, James Cronin, Val Fitch, and Rene Turlay, working at Brookhaven National Laboratory, New York, discovered that occasionally the debris from the long-lived mode consisted of two pions instead of three. Theory implies that this would be impossible if CP symmetry were valid; the $+1$ and -1 book-keeping would forbid it. The implication had to be that the bookkeeping fails: CP is *not* an exact symmetry of Nature. This means that in any beam of kaons, there will eventually be an excess of "matter" (the $K^0[d\bar{s}]$) over "antimatter" (the $\bar{K}^0[\bar{d}s]$) at the level of 1 part in 300. This is very tiny but nonetheless a real and highly significant discrimination between matter and antimatter.

The immediate questions that it raised were these. First: what force makes this happen? Is it unique to strange mesons, or are there other examples? If we can find other examples, could comparison with the K meson give enough clues to solve the mystery of how matter dominates over antimatter in the universe at large? The answers to the first two of these questions are now known. This is how the solution to the puzzle was found.

BETA DECAYS AMONG THE GENERATIONS

We saw (p. 124) how the (u, d) flavours are siblings in the sense of beta-decay, where the emission or absorbtion of a W^+ or W^- links $u \rightleftharpoons d$. From the world of charm we know that (c, s) are also in that the W radiation causes $c \rightleftharpoons s$. Once charm had been discovered, it was found that the propensity for $c \rightleftharpoons s$ in such processes was the same as that for $u \rightleftharpoons d$. This confirmed the idea that the flavours of quarks come in pairs, known as generations; and that but for their different masses, one generation appears to be indistinguishable from the other. Even more remarkable was that this universal behaviour happens for the generations of leptons, too. The analogous transitions $e^- \rightleftharpoons \nu_e$ and $\mu^- \rightleftharpoons \nu_\mu$ have the same properties as one another and as the quarks. The strengths of the lepton processes are the same as those of the quarks to an accuracy of better than 4%, or "one part in 25."

Here we see Nature giving a clear message that quarks and leptons are somehow profoundly related to one another. But even more remarkable was the observation by the Italian physicist Nicola Cabibbo in 1964 that the "one part in 25" was actually a real discrepancy! This led him to a theory that, in modern language, crosses the generations and is the seed for Kobayashi and Maskawa's theory of CP violation.

Today we know of the bottom and top flavours, which form the third generation, and that $t \rightleftharpoons b$ appears also to follow the same universality as the other two pairs. However, the 30 years of experiments since the discovery of charm have produced much more precise data than were available then. As a result, it is now clear that there are small but definite deviations from that apparent universality. The "one part in 25" is a real deviation, and there are others that show up when measurements are even more precise. Remarkably, all of these turn out to be linked together in a way that we can illustrate using nothing more sophisticated than the geometry of right-angle triangles and Pythagaros' theorem, and reveal that there is indeed a profound universality at work.

We saw (p. 129) how the transmutation $c \rightleftharpoons s$ can cause the lightest charm meson $D(c\bar{u})$ to decay, converting into a strange $K(s\bar{u})$ meson. If the strange quark were stable, then the K as the lightest meson containing a strange quark would be stable also. However, strange particles can decay; the weak force can destroy a strange quark, converting it to an up quark. The process $s \rightleftharpoons u$ is entirely analogous to $d \rightleftharpoons u$ but for one feature: it is about 25 times more feeble. The kaon-decay shows that there is "leakage" between the pairs in different generations; soon after the discovery of charm, it was found that there is an analogous leakage involving $c \rightleftharpoons d$ and that this too is enfeebled by about a factor of 25.

Thus the summary so far is that the strength of the weak force when acting within either one of the quark generations is (to within 1 part in 25) identical to that when acting on the leptons: $e^- \rightleftharpoons \nu$; however its strength is only about 1/25 as powerful when leaking between one pair and the other, $c \rightleftharpoons d$ and $u \rightleftharpoons s$.

There is a tantalising relation among these strengths. Suppose that we compare everything to the "natural" strength as typified by the leptons ($e^- \rightleftharpoons \nu$). The effective strength when leaking between generations of quarks is then $\sim 1/25$ of this. What Cabibbo had done was to take the "one part in 25" discrepancy as real and assume that the true strength between pairs of the same generation is therefore essentially $24/25$ relative to that of the leptons. This inspired him to the following insight into the nature of the weak interaction acting on quarks and leptons. It is as if a lepton has only one way to decay, whereas a quark can choose one of two paths, with relative chances $A^2 = 1/25$ and $1 - A^2 = 24/25$, the sum of the two paths being the same as that for the lepton.

Today we know that this is true to better than one part in a thousand. This one part in a thousand is itself a real deviation from Cabibbo's original theory, and is due to the effects of the third generation, which was utterly unknown in 1964. Such subtle effects arising from the third generation are what Kobayashi and Maskawa realised could affect the kaons and give rise to the breakdown of *CP* symmetry.

Let us start by ignoring such trifling effects and imagine a world of just two generations for which Cabibbo's theory would be enough. We can illustrate his accounting scheme with a simple piece of trigonometry, based on Pythagaros' theorem that the square on the hypotenuse of a right-angle triangle has the same area as the sum of the areas in the squares on the other two sides. Suppose the square on the horizontal side is 24/25 in some units, and that on the vertical is 1/25; then that on the hypotenuse is 1.

If the chance that something will happen varies between 0, meaning it definitely does not occur, and 1, meaning that it definitely does occur, then the 24/25 and 1/25 represent the relative chances that the beta-decay takes place within a generation (horizontal side) or leaks between (vertical side). So the whole accounting boils down to knowing the magnitude of the angle between the horizontal and hypotenuse. This is known as the Cabibbo angle. I will denote it θ_{12}, the subscript reminding us that it summarises the relative chances for the transitions within generations 1 and 2. We can draw the whole thing on a plane such as this sheet of paper (Figure 11.3)

Beta transition within a generation is thus represented by the horizontal length $A_H \equiv \cos\theta_{12}$ and that between generations by the vertical $A_V \equiv \sin\theta_{12}$. Each of

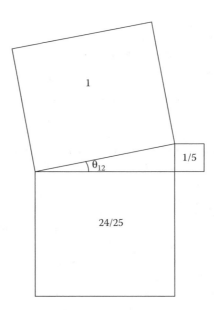

FIGURE 11.3 One Triangle Accounts for two generations. A horizontal line and a vertical line represent the amplitudes for a situation with two possible outcomes. The area of the square on each line is then proportional to their relative probabilities, and the square on the hypotenuse is then the sum of these, which is the total probability that something happens. As something definitely happens, we set the length of the hypotenuse to be unity. The set of all possible right-angle triangles of this sort form a circle with radius unity. The angle θ is all that we need to determine how the chance was shared in any particular case.

these lengths is known as an "amplitude" and it is the area of the square, given by the amplitude squared, that is measure of the chance or probability. So for the case of the quarks, it is traditional to say that $\cos\theta_{12}$ is the amplitude for $u \rightleftharpoons d$ and that $\sin\theta_{12}$ is the amplitude for $u \rightleftharpoons s$. If there were just two pairs linked in this way, the probabilities for $c \rightleftharpoons s(d)$ and $u \rightleftharpoons d(s)$ would be the same but their amplitudes would differ. It is traditional to write the four amplitudes in a matrix:

$$\begin{pmatrix} A(ud) & A(us) \\ A(cd) & A(cs) \end{pmatrix} = \begin{pmatrix} \cos\theta_{12} & \sin\theta_{12} \\ -\sin\theta_{12} & \cos\theta_{12} \end{pmatrix}$$

We could write the same matrix representation for the corresponding antiquarks. What is good for the quarks of matter is equally good for the antiquarks of antimatter in this example: the definition of θ_{12} and its empirical magnitude are the same.

The usefulness in this way of keeping account is most apparent when we have all three generations (pairs) involved (see Figure 11.4 on *universality*). The accounting for any pair of generations can be done by analogy with Cabibbo's angle. Weak interaction transitions between generations 1 and 3 can be described by a right-angle triangle with angle θ_{13} instead of θ_{12}. Suppose for simplicity that this second triangle

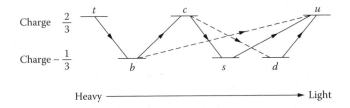

FIGURE 11.4 Universality. These matrix accounting methods for the way that the various quarks behave assume that the $t \rightarrow b$ has the same strength as the $u \rightarrow d$ and s or the $c \rightarrow s$ and d of the first two generations. This is itself profound. The fact that quarks feel a universal strength of the weak force, irrespective of what flavour or generation they are, suggests that the corresponding flavours in different generations are exact copies of one another, at least as far as weak interactions are concerned. Their electric charges are the same and their strong colour effects are also. Thus, apart from their masses, the corresponding quarks (u, c, t or d, s, b) are effectively identical to one another. This is a further hint of a deep unity underlying matter at its quark level. Similar remarks appear to hold for the leptons, too. It seems that there is a profound symmetry among them, broken by mass alone. Each downward arrow represents a decay emitting $e^{+}\nu$; each upward arrow emits $e^{-}\bar{\nu}$. Some less probable paths are also shown with dotted arrows.

is drawn in the same plane as the first so that we can draw them both on this sheet of paper (Figure 11.5). We now have all we need to keep the accounts.

Let us illustrate the idea by summarising the possibilities for the u flavour, which can link to any of d, s, b. When only d, s were involved, the chance for the beta transition from u to d was $cos^{2}\theta_{12}$ and that to s was $sin^{2}\theta_{12}$, their sum equalling unity. However, as there is also the chance to go to b, the total chance of going to d and s must now be less than unity. The chance of going to b is proportional to (the square of) $sin\theta_{13}$ and the three amplitudes are given by the three highlighted sides of the two conjoined triangles in Figure 11.5. The total probability, unity, is now the (square on the) hypotenuse of the upper triangle. Trigonometry shows that the amplitudes (lengths) corresponding to the individual transitions are then

$$(ub) = sin\theta_{13}; (us) = cos\theta_{13}sin\theta_{12}; (ud) = cos\theta_{13}cos\theta_{12}$$

Squaring each of these and adding them confirms that the sum total chance equals unity when all possible routes are allowed.

This simple piece of trigonometry summarises almost all of the accounting in the CKM scheme. Fundamental properties of quantum mechanics imply that the same angles work in this plane for quarks as for antiquarks. So there is no asymmetry between matter and antimatter in this. This is the one ingredient that our example does not describe. What is missing?

In this illustration (Figure 11.5a) we have drawn the two triangles in the same plane, but there is no general reason why they should be restricted like this. We have three dimensions to play with and the triangle containing the angle θ_{13} could rise out of the paper by some angle, let us call it δ (Figure 11.5b, c). If $\delta = 90°$, the triangle would be perpendicular to the page and pointing up at you, while $\delta = -90°$ would

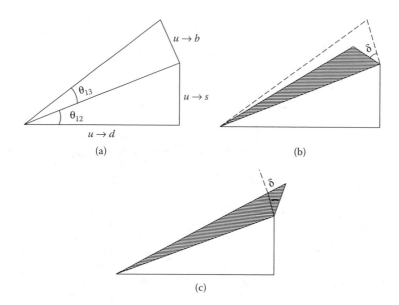

FIGURE 11.5 Two triangles and the accounts for three generations. (a) The accounts for generations 1 and 2 involved a single triangle with angle θ_{12}. This forms the lower triangle in the figure. Erect a right-angle triangle, with base along the hypotenuse of the first, and with angle θ_{13} between the new hypotenuse and the old. The bold sides of the triangles represent the amplitudes for transitions of (say) the u into, respectively, d (horizontal line), s (vertical line), and b (sloping line). (b) The effect of allowing the upper triangle to be oriented relative to the plane of the page introduces a further angle, δ. The projection of the sloping line onto the page is then proportional to $\cos\delta$ and its vertical distance out of the page is $\sin\delta$. (c) The triangles corresponding to the transitions of antiquarks, $\bar{u} \rightarrow \bar{d}, \bar{s}, \bar{b}$, are the same but with the upper triangle reflected relative to the plane of the page. This "phase" angle δ introduces a difference between quarks and antiquarks when weak transitions that cross the generations take place.

correspond to it being vertically down into the page. In general, any value for δ could be imagined; the special case $\delta = 0$ corresponds to it lying flat on the page, which was what we originally described.

Quantum mechanics implies that the plane of the page corresponds to an effective matter–antimatter mirror such that if the angle of orientation of the triangle is $+\delta$ for quarks, it is $-\delta$ for antiquarks. It turns out that this also corresponds to what would happen if one could imagine running the process backwards in time (we will apply the idea to the process where one starts with K^0 and ends with $\bar{K}^0$; the reverse process would then be $\bar{K}^0$ turning into K^0). This leads to some subtle potential differences between matter and antimatter in the weak interactions. We have gone as far as we can with this pictorial approach. To keep the accounts, we have to resort to matrices (see Table 8.2, Figure 11.5 and Appendix 11A). The angles $\theta_{12,13,23}$ are the Cabbibo angles and the δ keeps account of the quark and antiquark. Multiplying the matrices corresponds to computing the compounded probabilities. To see why this is important, let us first see what happened with the kaons and how the three generations may be playing a subtle role behind the scenes.

MATTER TO ANTIMATTER

The weak interaction can change a K^0 into a $\bar{K}^0$ or vice versa. It is a two-step process (see Figure 11.6). In a world of two generations, the chance of K^0 into a $\bar{K}^0$ is the same as the reverse and thus no asymmetry results. However with three generations there is a (small) chance for the intermediate stage to involve the third generation in the form of the massive top quark (this requires borrowing the energy associated with the top mass for a nugatory moment, as allowed by the quantum rules). In the imagery of Figure 11.5(b,c), one way involves the upper triangle being up out of the page and the reverse involves it pointing down into the page. In the more formal mathematics of the matrices (Table 8.2), this involves the i in Appendix 11a and the angle δ, which encodes a small asymmetry between one way and the reverse. Ultimately this leads to the asymmetries that have been observed under the banner of *CP* violation.

Bottom quarks are in effect heavier versions of strange quarks, and there will be bottom analogues of the strange kaon. Thus, in place of the "matter" $K^0[d\bar{s}]$ and "antimatter" $\bar{K}^0[\bar{d}s]$, we have their bottom counterparts: the B^0 and $\bar{B}^0$ made of $d\bar{b}$ and $\bar{d}b$. As was the case for the kaons, we can imagine two neutral Bs:

$$B^0_S \equiv B^0 + \bar{B}^0; B^0_L \equiv B^0 - \bar{B}^0$$

As above, the weak interaction can change a B^0 into a $\bar{B}^0$ or vice versa. According to the CKM theory, this should be a more dramatic effect than for the kaon. Essentially the difference comes because, for the kaon, the critical role of the "out-of-the-page" triangle for the the third generation came at a price: the kaon contains quarks of the first and second generations for which any transition to the third generation is a "leakage." This is in marked contrast to the bottom meson, made of $d\bar{b}$ or $\bar{d}b$, for which the intermediate stage involving the top flavour is now favoured: the B contains a b quark whose preferred transmutation by the weak interaction is into t. Having taken this step, the t is "forced" to convert back to d or b in order to return to the B, $\bar{B}$ meson (Figure 11.7). The latter step contains the factor i and leads to the asymmetries that we have been discussing.

The end result is that *CP* violation should occur here, too, but at a much more dramatic level than happens for the kaons. The strategy for investigating this is to make billions of these ephemeral B particles and their $\bar{B}$ antiparticle counterparts, and to study them in detail.

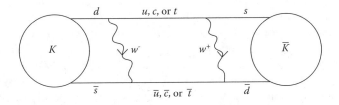

FIGURE 11.6 *K* turns to $\bar{K}$ by a two-step process. A $K^0(d\bar{s})$ converts into $K^0(s\bar{d})$ by the weak interaction acting twice, which is represented by the wiggly lines marked W^+ and W^-, respectively.

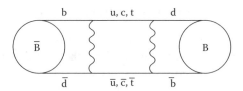

FIGURE 11.7 $B^0 - \bar{B}^0$ conversion. A $\bar{B}^0(b\bar{d})$ conversion into $B^0(d\bar{b})$ involves similar diagrams to the K^0 in Figure 11.6. The affinity of $t \rightarrow b$ enhances this B^0 case relative to the K^0.

The decays of Z^0 at LEP produced a large number but the most effective route has been to design customised "B-factories" where beams of electrons and positrons collide at energies tuned to a total of about 10 GeV, such that the production of bottom-flavoured hadrons is especially favoured. This has been done in Japan at "Belle" and at Stanford in an experiment known as "BaBar." One of the novel features is that the electron and positron beams have different energies, unlike other places, such as LEP, where they collide head-on at the same speed. The effect of this asymmetric configuration is that the B and $\bar{B}$ are produced with considerable momentum in the laboratory. This has two advantages for their study. One is that they fly away from their point of production, which makes them easier to detect; the other is a result of time dilation in Einstein's relativity. This has the effect that the faster a particle is travelling through space, the slower time elapses for it. Hence the B and $\bar{B}$ produced in the asymmetric e^+e^- collisions at the B-factories have their clocks effectively slowed; their elongated lifetime in motion enables them to be studied more precisely.

The accelerators were completed in 1999 and, after initial testing, began to collect data. To get definitive results requires creating and studying vast numbers of the bottom particles. It is like tossing a coin: chance might make it come up heads five or even ten times in a row; but if this continues to happen, then something is special about the coin. So it is with the study of these ephemeral particles. They live for less than the blink of an eye and it is what remains after they die, their fossil relics if you like, that has to be decoded. One needs to have huge numbers of such fossils in order to tell if any differences are real or the result of chance.

There are many different decay modes of B and $\bar{B}$ that can be studied. Among them is a particular species, $B(\bar{B}) \rightarrow \psi K_s$, known as "psi-K-short" events. The relative ease with which experimentalists could identify the ψ and the K_s, together with the theoretical assessment of this mode, suggested that this would be a measurable indicator of a difference between bottom-matter and bottom-antimatter.

By the end of the year 2000, definite hints of such a difference were being seen in these "psi-K-shorts," though it was not until the summer of 2001 that the results emerging from the Californian and the Japanese experiments were finally in agreement. By 2004 it became clear that the ψK_s data showed a large difference between B and $\bar{B}$, in accord with what had been predicted. Other decay pathways are being examined to see if a common story emerges, or whether more subtle phenomena will be revealed. At the LHC from 2007, B particles and antiparticles will be produced in quantites far exceeding those at the electron-positron B-factories. A dedicated experiment known as "LHC-b" will be investigating the properties of *B* mesons as part of this quest.

It seems likely that the origin of *CP* violation for strange and bottom flavours is on the threshold of being solved. While this shows that there is an asymmetry between *s* and also *b* flavours of matter and antimatter, it still leaves the origin of the bulk asymmetry between "conventional" matter and antimatter in the material universe as an unsolved puzzle. One possibility is that parity, *CP* violation, and the asymmetry between matter and antimatter are examples of symetries that have been "spontaneously broken," or "hidden." This concept is described further in Chapter 13.

APPENDIX 11A: CKM AND BOTTOM

We know of *three* generations: bottom quarks (*b*) are in effect heavier versions of strange and down quarks; top quarks (*t*) are likewise heavier versions of charm and up. A single angle θ_{12} was sufficient to do the accounting when there were only two pairs; for three generations, there are three rather than two possible routes for each of the (*t*, *c*, *u*) to link to any of the (*b*, *s*, *d*). Thus there are more possible routes to keep track of, but by treating the three pairs two at a time, we can do it.

Consider the (*ud*) pair. They can couple with the (*cs*) pair and for that accounting use the angle θ_{12}, or they could couple with the (*tb*) pair, for which we can do the accounting by a similar angle, θ_{13}. Finally, there is also the possible connection between the pairs (*cs*) and (*tb*) and this we can account by the angle θ_{23}.

The matrix for two generations is the 2 × 2 form in the main text:

$$\begin{pmatrix} u \\ c \end{pmatrix} = \begin{pmatrix} cos\theta_{12} & sin\theta_{12} \\ -sin\theta_{12} & cos\theta_{12} \end{pmatrix} \begin{pmatrix} d \\ s \end{pmatrix}$$

(refer to Table 8.2 to recall the rules for multiplying matrices). The elements in the rows link the *d*, *s* to the *u*, *c*. We can include a third generation (*b*, *t*), assuming for the moment that it is blind to the first two, (i.e., that $b \rightleftharpoons t$ only) by simply adding a third row and column, putting zeroes everywhere "to keep the third set apart from the first two" and a one at the bottom "so the beta transition for members of the third set stays in the third set"

$$\begin{pmatrix} u \\ c \\ t \end{pmatrix} = \begin{pmatrix} cos\theta_{12} & sin\theta_{12} & 0 \\ -sin\theta_{12} & cos\theta_{12} & 0 \\ 0 & 0 & 1 \end{pmatrix} \begin{pmatrix} d \\ s \\ b \end{pmatrix}$$

In reality, the (*u*, *d*) pair can link not only to *c*,*s* with angle θ_{12}, but also to *t*, *b* with angle θ_{13}. We can account for this connection of generations 1 and 3, ignoring generation 2 this time, by means of a matrix similar to that above but with the elements in rows and columns 1 and 3 in place of 1 and 2. Of course, the magnitude of θ_{13} and θ_{12} need not be the same so we will write the 3 × 3 CKM matrix for generations 1 and 3 as:

$$\begin{pmatrix} cos\theta_{13} & 0 & sin\theta_{13} \\ 0 & 1 & 0 \\ -sin\theta_{13} & 0 & cos\theta_{13} \end{pmatrix}$$

The full accounting for generation 1, when they can connect to either generation 2 or 3, involves multiplying these two matrices together:

$$\begin{pmatrix} cos\theta_{13} & 0 & sin\theta_{13} \\ 0 & 1 & 0 \\ -sin\theta_{13} & 0 & cos\theta_{13} \end{pmatrix} \times \begin{pmatrix} cos\theta_{12} & sin\theta_{12} & 0 \\ -sin\theta_{12} & cos\theta_{12} & 0 \\ 0 & 0 & 1 \end{pmatrix} =$$

$$\begin{pmatrix} cos\theta_{13}cos\theta_{12} & cos\theta_{13}sin\theta_{12} & sin\theta_{13} \\ -sin\theta_{12} & cos\theta_{12} & 0 \\ -sin\theta_{13}cos\theta_{12} & -sin\theta_{13}sin\theta_{12} & cos\theta_{13} \end{pmatrix}$$

The top row of the resulting matrix gives the amplitude for connecting u with $d, s,$ or b to be, respectively:

$$A(ud) = cos\theta_{13}cos\theta_{12}; \quad A(us) = cos\theta_{13}sin\theta_{12}; \quad A(ub) = sin\theta_{13}$$

which was illustrated by the triangles in Figure 11.5.

The accounting for generations 2 and 3 is only complete when we include the third analogous matrix:

$$\begin{pmatrix} 1 & 0 & 0 \\ 0 & cos\theta_{23} & sin\theta_{23} \\ 0 & -sin\theta_{23} & cos\theta_{23} \end{pmatrix}$$

and multiply this by the result of (12) × (13) above.

This is almost everything except for the extra feature of needing to include the δ rotation angle. It has become accepted practice to incorporate this with the matrix involving θ_{13}. The accounting requires that $sin\theta_{13}$ is multiplied by $cos\delta \pm isin\delta$ (where $i^2 = -1$) such that the (13) matrix is actually

$$\begin{pmatrix} cos\theta_{13} & 0 & sin\theta_{13}(cos\delta - isin\delta) \\ 0 & 1 & 0 \\ -sin\theta_{13}(cos\delta + isin\delta) & 0 & cos\theta_{13} \end{pmatrix}$$

The total accounting is now complete when the three matrices, including the complex numbers, are multiplied together. This is messy but thankfully experiment shows that the angles θ are very small and that an excellent approximation to the full answer can be written by the following CKM matrix (the plus and minus in the upper or lower signs in the $\pm$ and $\mp$ referring to quarks and antiquarks, respectively):

$$\begin{pmatrix} u \\ c \\ t \end{pmatrix} = \begin{pmatrix} 1 - \frac{\theta^2}{2} & \theta & A\theta^3(x \mp iy) \\ -\theta & 1 - \frac{\theta^2}{2} & A\theta^2 \\ A\theta^3(1 - x \pm iy) & -A\theta^2 & 1 \end{pmatrix} \begin{pmatrix} d \\ s \\ b \end{pmatrix}$$

12 Neutrinos

The basic particles that comprise matter on Earth are the *quarks* (up and down varieties to make protons and neutrons in atomic nuclei), the *electron*, and one other, the *neutrino*. The *neutrino* has been likened to an electron without any electric charge. This is a useful mnemonic to place it in the scheme of things, but do not regard it as more than that. Produced in beta-decay along with an electron, these two are siblings much as are the up and down quarks (see Figure 9.1 for a reminder of how these four occur in this fundamental nuclear transmutation). But neutrinos play a much richer role in Nature, the full implications of which are still to be elucidated.

The neutrino is one of the most pervasive varieties of particle in the universe, yet it is also one of the most elusive. It has no electric charge, has long been thought to have no mass and to corkscrew through space at the speed of light, passing through the Earth as a bullet through a bank of fog.

However, during the final years of the 20th century, evidence accumulated that neutrinos behaved as if they might have masses after all. First hints came by capturing occasional neutrinos from the hordes that the Sun emits; then anomalies showed up in neutrinos arriving in cosmic rays, too. In the past few years, proof has emerged and with it a whole new set of challenges to our understanding of the Standard Model.

First we should recall some of the neutrino story.

When Pauli first proposed the neutrino's existence, he was so certain that it could not be found that he wagered a case of champagne as incentive. Neutrinos interact so weakly with other matter that they are very difficult to detect. An illustration of this is the fact that a neutrino produced in beta-decay could pass clean through the Earth as if we were empty space. Thus neutrinos produced in the sun shine down on us by day and up through our beds by night. It is often said in popular accounts that a neutrino could pass through light years of lead without interacting; while this can be true, it begs the question of how we have managed to tease out evidence of its reality and to measure its properties, even to use it as a tool for other investigations.

The answer in part relies on the principle that is familiar in national lotteries: vast numbers of people take part and although the odds are that you or I will miss the jackpot, some random individual hits it. Thus it was with the discovery of the neutrino. When radioactive material decays at the Savannah River nuclear reactor in the United States, vast numbers of neutrinos are produced. Over a million million of them pass through each square centimetre of its surface per second. This enormous concentration gave the possibility that if many tons of a suitable material were placed alongside, then occasionally a neutrino would interact with it. In 1956, Frederick Reines and Clyde Cowan observed the tell-tale signs of neutrinos having hit atoms in their detector — the neutrino produced electrically charged particles that were easily seen. (This is similar in spirit to the way that the neutron had revealed itself by ejecting protons from paraffin wax (p. 33) Pauli paid up the champagne that he had promised a quarter of a century earlier.

Neutrinos are produced in radioactivity, in nuclear reactors, and in the Sun; such neutrinos have quite low energies. This contrasts with cosmic rays and particle accelerators, which also can produce neutrinos but with very high energies. The propensity for neutrinos to interact with other particles grows in proportion to the neutrino's energy, thus making intense beams of *high-energy* neutrinos increase the chance of capturing an occasional one. This has enabled quantitative science to be done using neutrinos. As a result we have learned a lot, not least that there are three distinct neutrino varieties, known as the electron-neutrino, muon-neutrino, and tau-neutrino (or *nu-e*, *nu-mu*, and *nu-tau* for short; in symbols v_e, v_μ, and v_τ) in recognition of their respective affinities for the electrically charged electron, muon, and tau (see Table 12.1). It is the relation among these that is now taxing the ingenuity of the theorists.

The fusion processes at the heart of the Sun emit neutrinos of the electron variety nu-e (v_e). Solar physicists understand the workings of the Sun very well, and from its temperature and other features have calculated that its fusion processes emit 2×10^{38} neutrinos every second. These neutrinos have only low energy and so have little propensity to react. The Sun is almost transparent to them and they stream out across space. Some of them head in our direction. As you read this, hundreds of billions of them are passing through you each second (see Table 12.2). With such numbers it is possible to capture occasional ones if the detector is sufficiently large (see Table 12.3). Experiments have been recording these neutrinos for over 20 years, and verified that the Sun is indeed powered by fusion, but the actual number of neutrinos detected was about two to three times smaller than the solar theories required. This became known as the "solar neutrino problem." Suspicion grew that something happens to the neutrinos en route.

One exciting suggestion has been that they were passing into the fifth dimension, effectively disappearing from our four-dimensional universe. (The idea that there are higher dimensions than those that our senses are aware of has been seriously suggested, see Table 12.4.) Even a trifling chance of such a disappearing act can have a noticeable effect during a journey of 150 million kilometres. and could be a unique way of revealing such properties of space-time. However, the possibility that the vanishing *nu-e* (v_e) is the first evidence for such ideas now seems unlikely, the explanation of the missing neutrinos being consistent with the phenomenon of 'neutrino oscillations' — the quantum switching between neutrino types, such that what starts out as one variety of neutrino can transmogrify into the other varieties along the way. Theory shows that neutrinos oscillate from one variety to another

---

Table 12.1 Three Varieties of Neutrino: part 1

The three varieties, — v_e, v_μ, and v_τ — are produced, respectively, in association with e, μ, and τ. When they interact with matter and convert neutron into proton, they turn back into e, μ, and τ. (For many years this was thought to be absolutely true, but today we know that there is a small chance of a neutrino produced as v_e to end up as v_μ or v_τ; this is due to the phenomenon of neutrino oscillations; see Figure 12.1.)

How do we know that there are only three varieties of neutrino? We don't. However, we are certain that there are only three varieties of (nearly) massless neutrinos. This comes from the measured lifetime of the Z^0 (Chapter 10).

Table 12.2 Neutrinos Everywhere

The fusion processes in the Sun emit 2×10^{38} neutrinos per second. These stream out in all directions. We are some 150 million kilometres from the Sun. If we drew an imaginary sphere with the sun at its centre and with that radius, its surface area would be $4\pi(1.5 \times 10^{11})^2$ m^2, which is $\sim 3 \times 10^{23}$ m^2. So there are some 60 billion neutrinos from the Sun passing through each square centimetre, of such a sphere, and hence the Earth, per second.

The rocks beneath our feet are radioactive and even our bodies are radioactive, due in particular to isotopes of calcium and potassium in our bones, such that we emit some 400 neutrinos per second. These travel out into the universe and can indeed travel for light years through lead, or effectively forever through space without hitting anything. This is as near to immortality as we can imagine.

Table 12.3 Neutrino Detectors

As any individual neutrino is very unlikely to interact, you need lots of them and a large detector so that a few may be caught. When a neutrino hits, it has a tendency to turn into an electrically charged particle such as an electron, which is easy to detect. So detecting a neutrino is a secondary process, based on recording its progeny.

It is possible for a neutrino to travel through a substance, such as water, faster than light does. This produces an electromagnetic shock wave known as Cerenkov radiation. This property is exploited in some neutrino detectors where the charged particles that the neutrino has produced travel at superluminal speed through the water. Then one records the Cerenkov radiation and infers the passage of the electron and hence the original neutrino.

At SuperKamiokande (p. 168) the detector consisted of a huge tank of pure water, 40 m diameter and 40 m high with 13000 "photmultiplier tubes" around its walls. These are very sensitive light detectors, like lightbulbs in reverse in that when a flash of light goes in, electric current comes out and is recorded by computer. The flash of light can be caused by a neutrino hitting protons in the water; a problem is that the light is more likely caused by other things and so a major part of the experiment involves sorting out the needles from the haystack.

Similar principles are used at SNO (p. 169).

Table 12.4 Higher Dimensions

One of the problems in incorporating gravity into a unified theory is its remarkable weakness, orders of magnitude more feeble than even the weak force. An idea is that there are dimensions beyond the three space and one time dimensions that we are normally aware of. Gravity alone among the forces leaks into and through the higher dimension(s), which leaves its effects diluted within our immediate experience. The other forces act only within our conventional four-dimensional space-time. The problem is how to test such an idea. It "explains" dark matter as the effect of galaxies existing in the higher dimension, whose light does not leak into our space-time but whose gravitational effects do. However, there are other theories of dark matter and this is not definitive. Particles disappearing into the fifth dimension would give rise to effects that would appear to violate energy and momentum conservation. Such tests are planned at the LHC — the powerful new accelerator at CERN — as some theories predict that the effects of higher dimensions could become apparent in collisions among particles in excess of 1 TeV.

The fusion processes in the centre of the Sun produce neutrinos of the v_e variety. For simplicity, let us imagine there are only two varieties, v_e and v_μ. They are identified by the way they behave in weak interactions, being produced in association with electron and muon, respectively. If we were able to measure the masses of neutrinos, we would find two distinct states; let us call them v_1 and v_2 with masses m_1 and m_2, respectively. It is possible that v_1 is the same as v_e, say, but, in general, a mixture is possible:

$$v_e = cos\theta(v_1) + sin\theta(v_2)$$
$$v_\mu = cos\theta(v_2) - sin\theta(v_1)$$

where θ is called the mixing angle. For further simplicity let us assume θ is 45° so that $cos\theta = sin\theta$ and the relation is

$$v_e = [(v_1) + (v_2)]/\sqrt{2} \qquad (12.1)$$
$$v_\mu = [(v_2) - (v_1)]/\sqrt{2} \qquad (12.2)$$

It is the states $v_{1,2}$ that have well-defined masses and determine the frequency of the probability waves that spread out across space. If the masses differ, the wavelengths of the two pieces differ, as in Figure 12.1. After awhile, the peaks and troughs of the two waves get out of step ("out of phase"). If they are completely out of phase, such that one is at its peak where the other is in its trough, then the sum will be nothing. But Equation (1) shows that the sum is the magnitude of v_e at this point; so this means that there is **no** v_e here! What started out as v_e at the source has vanished.

Now look at the expression for v_μ. The minus sign between the two terms means that if we take a peak (of wave 1) and **remove** a trough, this is like adding extra strength to the original peak. This implies that there is a powerful v_μ presence at this point. So the full story is that what set out as v_e has "oscillated" into a v_μ. Continue onwards for the same distance that the waves have already come and they will have oscillated back again to 100% v_e and vanishing v_μ.

The probability that what set out as v_e is still v_e varies as a function of the distance travelled from the source. As the probability for v_e falls, that for the v_μ grows, and vice versa. We know of three varieties of neutrino and so the mixing could involve all three. Disentangling the details of these is one of the current goals in particle physics.

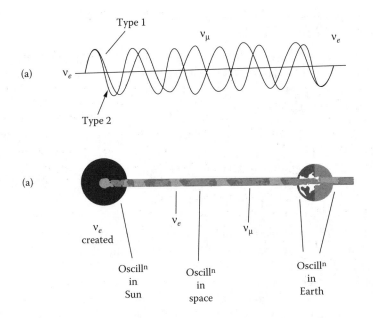

FIGURE 12.1 Neutrino Oscillations. (a) The upper figure shows a v_e composed of two mass states $v_{1,2}$ which oscillate at different rates. When they are out of phase, in this example, it corresponds to only v_μ being present, the v_e having vanished. (b) In the lower figure, a v_e formed in the Sun may oscillate back and forth during its transit through the Sun, through space, and even through the Earth (at night). A comparison of the v_e intensity by day and by night could reveal the effect of their passage through the Earth.

only if they have different masses. If physical neutrinos are a mixture of two or more underlying quantum states with different masses, then for a given energy, the heavier will travel more slowly than the lighter. The quantum waves of these underlying states have different frequencies and so swell and fade back and forth into one another. The smaller their mass difference, the more alike their speeds, the slower their oscillation and the longer the distance needed for any effect to be detectable. (See Figure 12.1 and accompanying text.)

As neutrino masses are themselves very small, any differences must also be tiny and oscillation lengths large. The experiments detecting neutrinos from the Sun were sensitive only to the (v_e) variety, so it was suggested that the shortfall might be because they had changed into (v_μ) or (v_τ) during their long journey. The radical idea that neutrinos can change their identity began to be taken very seriously when a similar phenomenon was discovered in cosmic rays.

High-energy cosmic rays hitting the atmosphere produce a cascade of secondary particles. These eventually lead to a shower of neutrinos at ground level, and even below ground. From experience with the content of cosmic rays and the products of their collisions in the upper atmosphere, it was calculated that there should be twice as many (v_μ) as (v_e). However, experiment found a significant shortfall of the (v_μ) variety; cosmic rays were revealing that v_μ disappeared, while the Sun suggested that (v_e) disappear. Neutrinos from the Sun have travelled millions of kilometres, and

Table 12.5 Mixing Angle

The rate of oscillation depends on the different masses; the strength of the swell depends also on the mixing. This is described by a "mixing angle." The standard model has six quarks and six leptons. The mixing angles for the quarks have been known for a long time. Quarks can be listed according to their masses, for example, the massive bottom quark, middling strange, and light down, or by their response to the weak force. If there was no mixing, the bottom and strange quarks would be stable; in reality, they decay slowly as a result of a finite, but small, mixing angle (the Cabibbo angle, Chapter 11). The magnitudes of the various mixing angles seem to be related to the relative masses of the different quarks, but a detailed explanation is still awaited.

The similarities between the leptons and quarks in the Standard Model have led to the idea that at energies far higher than physics can yet attain, a profound unification among the particles and forces will be revealed. The fact that the mixing angles are small in the quark sector had led to a suspicion that they would be small also for the leptons. However, when the first hints of ν_μ oscillations in the cosmic ray data emerged, they implied that the relevant mixing angle is large. In light of this, and as the data at that stage were not that conclusive anyway, many theorists dismissed them. However, data from SuperKamiokande convinced even hardened sceptics: mixing angles for neutrinos are large.

This does not rule out unification — there are too many hints that there is some profound linkage between quarks and leptons — but the details still remain to be worked out. Measurement of neutrino masses and mixing angles may eventually give important clues as to the nature of unification and the physics that lies beyond the Standard Model.

those from cosmic rays up to 10,000 km, hence the propensity for oscillation en route and the "disappearance" of ν_e and ν_μ, respectively.

The first hints of oscillations did not fit easily with some theorists' expectations (see Table 12.5). This changed as a result of two new lines of experiments. One involved the detection of oscillations in customised beams of neutrinos produced at high-energy accelerators; the other was the dramatic solution of the solar neutrino problem. To have any chance of detecting oscillations in beams of neutrinos produced at particle accelerators, it is necessary to place the detector far away from the accelerator, at distances of hundreds of kilometres, perhaps even in another country! June 1999 saw the start of the world's first 'long-baseline' experiment with neutrinos from an accelerator. In Japan, a neutrino beam created at the KEK laboratory travels 250 km westwards under the Hida Sammyaku (the Japanese 'Alps') to the SuperKamiokande detector (known as "Super-K").

This huge detector was constructed principally to study neutrinos travelling 150 million kilometres from the Sun, but had also provided some of the most persuasive evidence for the oscillation of neutrinos made in cosmic rays, even being able to distinguish between those produced overhead from those produced 13,000 km away in the atmosphere on the opposite side of the Earth! With the beam from KEK, it began studying neutrino *baselines* of 250 km, thereby covering a wide range of possibilities for establishing the details of oscillations.

The KEK produces a billion ν_μ each second in their beam. Using the GPS system to do precision positioning, the scientists were able to identify each pulse of neutrinos emerging from the accelerator and correlate it with the pulses arriving at the detector. After travelling 250 km, the beam has spread out, and the chance of neutrinos hitting protons in the water is also tiny, so only a few hits are expected. Even so, the first results found a shortfall, confirming the results from the cosmic rays.

A second long-baseline experiment, known as MINOS (Main Injector Neutrino Oscillation Search), consists of two detectors: a small one close to the source of neutrinos produced at the Main Injector accelerator at Fermilab near Chicago, and a large one 710 km away, in the Soudan Mine in Minnesota.

Like the Japanese experiment, MINOS looks for a reduction in the numbers of ν_μ that reach the distant detectors. An alternative technique is to look for the appearance of ν_τ in the ν_μ beams, by detecting the τ particles they produce in their rare interactions. This is being investigated in an experiment directing a neutrino beam from CERN towards the Gran Sasso Laboratory, which is about 730 km away, under the Gran Sasso massif northwest of Rome.

The above experiments all involve ν_μ. While they were being prepared, dramatic discoveries were made about ν_e: the solar neutrino puzzle was solved.

As we saw earlier, the fusion processes in the Sun produce ν_e, and these had been found to be in short supply, the popular suggestion being that if the three varieties of neutrino oscillated back and forth from one form to another during their flight, then it could be natural that on average only one third would end up as ν_e by the time they reached Earth. As the Earthly detectors were only sensitive to ν_e, it had been suggested that this might explain the apparent shortfall.

The challenge was to design a detector that would be sensitive not just to the electron-neutrinos but to all three varieties. In 1990, a 100-member team began to build such a detector 2 km underground, where it is protected from all cosmic rays other than neutrinos, in Sudbury, Ontario (Canada). So was born SNO, the Sudbury Neutrino Observatory. It would reveal all neutrinos that arrive from the Sun, whatever tricks to hide they might have adopted. The key was to use ultrapure heavy water.

The proton that is the nucleus of hydrogen as found in ordinary water is accompanied by a single neutron in heavy hydrogen and heavy water. The proton and neutron in concert enable all the varieties of neutrinos to be exposed. Electron-neutrinos are revealed when they hit the neutron and convert it into a proton. By contrast, all three varieties of neutrino could hit either the neutron or proton and bounce off, leaving them unchanged. It is the recoil of the neutron and proton that give the neutrinos away in this case. By comparing the number of neutron and proton trails, the scientists can compute both the total neutrino flux and the fractional contribution of the electron-neutrino variety.

SNO is the size of a ten-story building and contains 1000 tonnes of ultra-pure heavy water enclosed in a 12-m diameter acrylic plastic vessel, which is, in turn, surrounded by ultra-pure ordinary water in a giant 34-m high cavity. When neutrinos are stopped or scattered by the heavy water, flashes of light may be emitted, which are detected by about 9600 light sensors.

It took 8 years to build and test SNO. At a detection rate of about one neutrino per hour, it took 4 years of dedication to get the first meaningful results. They found that the *total* number of neutrinos arriving here — the electron, muon, and tau varieties — agrees with the number expected, based on the latest sophisticated models of the solar core. So for the first time we had direct quantitative evidence that the Sun, and stars like it, are indeed powered by thermonuclear fusion.

This qualifies as one of the great moments in experimental science. But there is more. They confirm that electron-neutrinos only number a third of the total, which shows unambiguously that electron-neutrinos emitted by the Sun have changed to muon- or tau-neutrinos before they reach Earth. This can only happen if the neutrinos — long thought to be massless particles — have different masses. What the magnitudes of those masses are is beyond current technology to say, but that will eventually be determined.

What does this imply for the Standard Model, for the future direction of physics, or even the Universe at large?

In the Standard Model, quarks and leptons have an intrinsic spin and as they travel can effectively spin like a left- or right-handed screw. They gain mass as a result of interacting with Higgs bosons. Any particle with a mass will travel slower than the speed of light, and so it is possible to overtake them. Suppose you saw such a particle rotating like a left-handed screw as it passed you. Now suppose that you moved so quickly that you could overtake it. As you look back on it, it will now appear to be departing from you, and to be spinning in the opposite sense, as a right-handed screw. In a relativistic theory, mass mixes left and right.

In the Standard Model, neutrinos are left-handed. The absence of right-handed neutrinos and the supposed masslessness of the neutrino were linked. Antineutrinos, by contrast, are right-handed.

What happens when neutrinos have mass? The left-handed neutrino can be affected in two ways. One is simply akin to what happens with other particles: there can be a right-handed piece analogous to the way that an electron can have left- or right-handed pieces. This is known as a "Dirac" mass, after Paul Dirac, creator of the original relativistic quantum theory of the electron with its implication for the existence of antiparticles, the positron.

However, there is a second possibility for neutrinos, known as a "Majorana" mass, named after Ettore Majorana, the Italian theorist who first recognised the possibility. The left-handed neutrino could convert into the right-handed antineutrino. This switch between particle and antiparticle cannot happen for an electron as the electron and positron have opposite electric charges and the conservation of electric charge forbids it. Neutrinos have no electric charge and no sacred principle such as this is known to forbid it.

What then would distinguish neutrino from antineutrino? Neutrinos can pick up charge when they hit nuclei, and turn into electrons, say; antineutrinos correspondingly convert into positrons. So if neutrinos have Majorana masses, there will occasionally be "wrong" processes where they convert to a positron rather than an electron in such interactions. There should also be a variety of rare possibilities that violate the conservation of lepton number (the accounting scheme that keeps track of the numbers and varieties of leptons that appear and disappear in particle creation and decays). Experiments are looking for examples of these rare processes.

If the Majorana mass is confirmed, there is the tantalising possibility that in addition to the presently known lightweight, left-handed neutrino, there could exist a very massive right-handed neutrino. In this theory, each of the three familiar neutrinos could spend a small amount of time as this supermassive particle, which would give the neutrinos their small masses, triflingly small because the quantum fluctuations are so rare. It is as if the scale of energy or mass of around 10^3 GeV, where the Higgs boson — bearer of mass — is predicted to be, is the geometric mean of the supermassive and ultra-light neutrinos. In the jargon, this is known as the *see-saw mechanism*.

The known neutrinos have masses, but it would probably take over 100,000 of them to add up to the mass of the electron, which is the lightest measured particle to date. This vast disparity is tantalising. If this ultra-light neutrino mass scale is indeed one side of a see-saw from an ultra-heavy scale, then we have our first glimpse of the ultra-high energy world that exists beyond the Standard Model. If such supermassive right-handed neutrinos exist, there are interesting questions about their possible role in cosmology, their mutual gravity being so large as to have helped seed the formation of galaxies or be part of the mysterious "dark matter" that seems to pervade our present universe.

So where does this leave the Standard Model? This is the theme of Chapter 13.

13 Beyond the Standard Model: GUTS, SUSY, and Higgs

The Standard Model is not the last word — it is essentially the summary of how matter and forces behave at energies so far explored. It is generally recognised as being an approximation to a richer theory whose full character will eventually be revealed at higher energies, such as those available at the Large Hadron Collider (LHC) at CERN.

The relation between the Standard Model and the more profound theory of reality, which will subsume it, can be compared to Newton's Theory of Gravity and Einstein's Theory of General Relativity. Newton's theory described all phenomena within its realm of application but is subsumed within Einstein's Theory of General Relativity, which reaches parts that Newton's theory cannot. So it is with the Standard Model. It describes phenomena from energies on the atomic scale of fractions of $1\,\text{eV}$ up to around 10^3 GeV, the highest at which terrestrial experiments can currently be made. It has unimpeachable success over a range of energy spanning more than 12 orders of magnitude. However, it has no explanation for the magnitudes of the masses and other parameters such as the strength of the forces, which appear so critical for the emergence of life. The answers lie beyond the Standard Model, in a richer deeper theory of which the Standard Model will one day be seen as an approximation.

Ten years of experiments at LEP tested the Standard Model to the extremes. This has enabled us to sharpen our predictions of how to produce the Higgs boson, which is theorised to be the source of mass for the basic particles and the W, Z bosons, and also has given hints that "supersymmetry" may be part of Nature's scheme. With the discovery that the enigmatic neutrinos do have masses, we may have gained a first glimpse of physics beyond our present energy horizons. However, the ultimate foundations of reality remain to be discovered, perhaps at the LHC.

Before describing these ideas, we must first review some immediate implications of the Standard Model, which have been recently confirmed and point towards the ultimate theory. The Standard Model is stimulated by patterns shared by the particles and the forces, which seem to be too clear-cut to be mere chance. We may be getting here the first glimpse of a profound unity in Nature that existed during the Big Bang epoch but which has become hidden during the expansion and cooling of the universe. It is only with the creation of local 'hot' conditions in particle collisions using high-energy accelerators that a glimpse of this one-time unity has been obtained.

COINCIDENCES?

PARTICLES

The fundamental particles of matter, quarks and leptons, all have spin $\frac{1}{2}$; obey the Pauli principle, which prevents more than one particle of a given kind from being in any state; show no evidence of internal structure; and show a left-handed preference in weak interactions. In earlier chapters we have seen that the leptons and also the quarks form pairs, which controls their response to the weak force. The W^+ and W^- of the weak force act on the lepton pairs in precisely the same way as on the quark pairs: as far as the weak interactions are concerned, leptons and quarks appear to be identical (at least when the mixing angles $\theta_{12,13,23}$ of Chapter 11 are taken into account). Are quarks and leptons therefore not two independent sets of particles, but instead related in some way? Is the fact that each lepton pair is accompanied by a quark pair the first hint, as with Mendeleev historically, of a yet deeper structure?

The second question is still open though their profound similarities hint that there is something special about this layer of the Cosmic Onion. This contrasts with the first question, where a possible answer has emerged as a result of insights into the nature of the forces. To set the scene for this we should first not forget the electromagnetic and strong forces since here the leptons and quarks seem to behave quite differently: perhaps the weak interaction similarity was a red herring.

The quarks have charges $\frac{2}{3}$ or $-\frac{1}{3}$, the leptons have -1 or 0, and so the strength of coupling to the electromagnetic force differs for each. However, it differs only in the overall scale of $\frac{2}{3} : -\frac{1}{3} : 1 : (0)$. Once these relative sizes and their different masses are allowed for, all electromagnetic properties are identical, notably the ratio of magnetic moment and electrical charge is the same for the electron and for the up and down quarks.

The place where a manifest difference appears is in strong interactions, which quarks respond to whereas leptons do not. The quantum chromodynamic (QCD) theory relates this to the fact that quarks carry colour, which leptons do not have. However, even this difference gives hints of a deeper unity: there appears to be a correlation between the existence of three colours and third fractional charges for quarks and the non-fractional charges of uncoloured leptons (see Table 13.1).

FORCES

Until 1970 the strong, weak, and electromagnetic forces seemed to be totally unrelated to one another, the strong force being a hundred times more powerful than the electromagnetic force; the weak force yet another thousand times more feeble; and different varieties of matter being affected by these forces in quite different ways. But our perspectives changed with development of the quantum chromodynamics and electroweak theories, and with the discovery that they differ in strength only at low energies or at nuclear distances — one small region of Nature's spectrum. At high energies and very short distances, all of their strengths may become the same.

Electromagnetism involves *one* (electrical) charge and quantum electrodynamics mathematically is a U(1) theory. The pairs of weak isospin that quarks and leptons form (at least in the left-handed universe) is a two-ness that is the source of an SU(2)

Table 13.1 The Puzzle of the Electron and Proton Electric Charge: Episode 2

On p. 111 we commented on the profound puzzle of why the proton and electron have the same magnitude of charge, though of opposite sign. The simple idea that the electron charge can arise by adding a discrete unit to a neutrino, while the proton's analogously follows from a fundamental neutron cannot be the whole story as we now know that the neutron is made of quarks. However, the idea can be applied to leptons and quarks, starting from the fundamental neutrino *in both cases* so long as we include colour.

Starting with the neutrino and removing one unit of charge gives the electron as before. If we add a unit of charge to the neutrino, we would not obtain a known fundamental particle (the e^+ is the antiparticle of the electron). However, suppose we admit the existence of three colours and paint this mythical positively charged particle any of three ways, sharing the electric charge accordingly. Then we will have red, yellow, or blue spin $1/2$ objects each with charge $1/3$. This is nearly a quark: the up quark of charge $2/3$ and the down quark of charge $-1/3$ add to this very value.

So far, electric charge and colour (strong force) have entered the scene. Now consider also the weak interaction. This connects v^0 to electron, and down quark to up, so the *difference* of up and down charges must equal that of the leptons. Starting from a charge of $1/3$ that is to be distributed between the up and down quarks only allows the solution: up $= +2/3$; down $= -1/3$. The electron and proton (uud) electric charges are then precisely balanced.

To obtain this result we have had to make two profound assumptions:

1. Quarks and leptons are intimately related: in a sense, quarks are leptons that have been given colour.
2. The strong (colour) and weak forces have to conspire with the electromagnetic force (charge).

This suggests that these forces are not independent of one another.

The strength of the force between two charges depends in part on what exists between them. Two electric charges interact by exchanging a photon. En route, this photon might fluctuate into an electron and positron which then return into another photon. Such a "quantum fluctuation" subtly alters the force between the original charged particles. The more possibilities there are for such fluctuations, such as a $\mu^-\mu^+$ or quark and antiquark, the greater is the effect. The ease with which these fluctuations occur depends on the amount of energy and momentum that is transferred between the two charges. The net result is that the strength of the force, which is summarised by the perceived charge that one particle sees from the other, varies as the energy and momentum vary. In this example, where all the charged particles in the quantum fluctuations are fermions — have spin $\hbar/2$ — the effective charge increases as energy grows. Thus the strength of α in QED grows with energy.

In QCD, two coloured quarks can exchange a gluon. This gluon can fluctuate into a quark and antiquark, which will tend to increase the effective charge, strengthening the force, by analogy with the QED example. However, the gluon can also fluctuate into a pair of gluons (see Figure 7.6). These are bosons and their net effect goes in the opposite direction to what occurs with fermions: the gluon fluctuations tend to *decrease* the effective charge, enfeebling the force. As the gluons occur in any of eight colours whereas quarks have only three, the gluon fluctuations dominate over the quarks and the total result is that the force becomes feebler with increasing energy. This is opposite to the case with QED.

At very high energies, the photons of QED can even fluctuate into the massive W^+W^-, which being bosons can begin to enfeeble the charge again. It is more complicated in this example though because at energies of hundreds of GeV the electromagnetic and weak SU(2) merge, with photon and Z^0 playing comparable roles in transmitting the forces, and quantum fluctuations involving both the W^+W^- and also Z^0. The total effect is that the combined electroweak force changes its strength and tends towards that of the QCD colour force at energies of around 10^{15} GeV. For this coincidence to be achieved precisely, it appears that the menu of known particles is insufficient. However, if SUSY particles occur at high masses, they will also contribute to the variation of the α with energy. It turns out that if the lightest supersymmetric particle is of the order of 1 TeV, or perhaps even a bit less, then the forces do indeed appear to focus to a common strength at very high energies. Thus searches for evidence of supersymmetry at the new, high-energy accelerators are a high priority.

---

theory similar to quantum electrodynamics. Combining these yields the SU(2) × U(*1*) theory of weak and electromagnetic phenomena. The *three* colours possessed by quarks generate an SU(3) theory of quark forces, quantum chromodynamics, similar again to quantum electrodynamics. Quarks feel the strong force because they carry this threefold colour (charge): leptons do not have colour and are blind to it. Thus there is a common principle at work, the essential difference being the one-, two-, or three-ness realised in the U(*1*), SU(2), or SU(3) theories. This seems too much of a coincidence, and suggests that the three theories may be related.

A further hint of unification comes from the relative strengths of the three forces at low energies. The photon, W, and Z bosons, and the gluons are the quantum bundles that transmit the forces described by the U(*1*), SU(2), and SU(3) theories, respectively. The strength of their coupling to the relevant particles that experience the forces (electrically charged; flavour and coloured particles, respectively) are given by numbers g_1, g_2, and g_3. (Conventionally $g^2/4\pi$ is denoted by α, which $= 1/137$ for electromagnetic interactions and is in the range of 1 to 1/10 for coloured quark interactions in the proton at present accelerator energies.) Although electromagnetic interactions are weaker than 'strong' interquark forces at present energies, we should remember the remarkable property of the SU(3) theory, noted on p. 132, that the interaction strength g_3 is not a constant but decreases as energy increases. Thus although strong at low energies (large distances), the interquark force weakens significantly at higher energies. This property is not peculiar to SU(3) but occurs for all such theories, the only difference between SU(2) and SU(3) being the rate at which the weakening occurs as energy increases. For U(*1*), the force is predicted to get stronger at higher energies (and precision measurements at LEP confirm that at the energies where the Z is produced, α has increased slightly to a value of about $1/128$). Hence, the electromagnetic force gets stronger at high energies and the strong force weakens. If we extrapolate this theory to the huge energy of $\sim 10^{15}$ GeV, it implies that all three forces turn out to have comparable strengths: $\alpha(QED)$ has risen from $1/137$ to about $1/40$, and $\alpha(QCD)$ has fallen from unity to a similar value (Figure 13.1), but supersymmetry (next section) seems to be important in doing this.

The forces are described by a common mathematics and have similar strengths at 10^{15} GeV. Under these conditions the idea of unification has some meaning; the unity is obscured at the cold low energies to which science had initially been restricted. It was only in the final quarter of the 20th century that accelerators were capable of colliding particles at energies exceeding 500 GeV, and it was only with the advent of these high energies that the changing strengths of the forces were measured. We are never likely to be able to collide individual particles at 10^{15} GeV in the laboratory and see the full glory of the unity of the forces, but if the hot Big Bang model of creation is right, then the head-on collisions at such energies would have been abundant in that epoch. The current belief, which has exciting consequences for cosmology, is that in the high temperatures of the hot Big Bang there was indeed a unity among the natural forces with consequent production and decay mechanisms for particles that have been long forgotten as the universe cooled. Some of these processes appear to have been crucial in generating the universe that we see today.

The idea that symmetrical behaviour occurs in phenomena at extreme temperatures but is obscured at low temperatures is familiar in many contexts. One such is the transition from liquid to crystalline solid. At high temperatures, liquids are isotropic, all directions being equivalent to one another. As the temperature drops, the liquid solidifies and may form crystals. These are not isotropic; the original full symmetry has been lost though well-defined symmetries, less than total isotropy, remain, which give crystals their characteristic shapes. Magnets yield a second example: at high temperatures the atomic spins are isotropically randomly ordered and no magnetism ensues, whereas at lower temperatures north and south magnetic poles occur. Thus the notion that the universe possessed an innate symmetry at extremely high temperatures, such as at the Big Bang, which was obscured as it cooled to the present day is quite in accord with other physical experience. While there is a general consensus that these hints suggest that our cold, structured universe is the frozen remnant of a hot, highly symmetric situation, the details and identity of the true unified theory, and the mechanism by which it gave rise to the disparate particles and forces today, are still hazy.

In this philosophy, leptons and quarks are the distinct remnants at low energy of some unified family of "leptoquarks" that exist(ed) at high energy. The price of this unification is that leptons and combinations of quarks can convert into one another — in particular, that a proton may then decay into a positron and gamma rays, for example. (See also Table 13.2.) This prediction continues to run against experiment as there is no convincing evidence that this occurs; protons appear to be stable with lifetimes in excess of 10^{32} years (!). The problem of uniting the forces and particles without at the same time implying unacceptable instability of the proton is currently an unresolved issue.

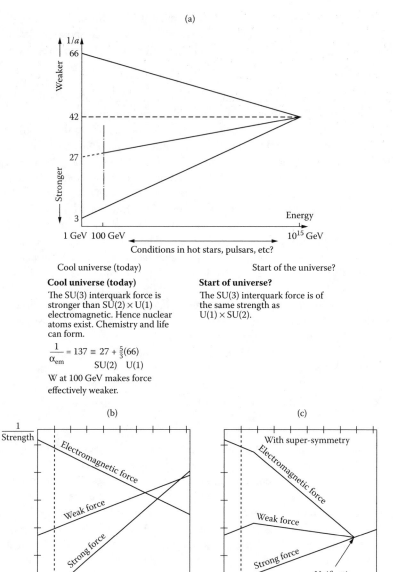

(a)

Cool universe (today)

The SU(3) interquark force is stronger than SU(2) × U(1) electromagnetic. Hence nuclear atoms exist. Chemistry and life can form.

$$\frac{1}{\alpha_{em}} = 137 \equiv 27 + \tfrac{5}{3}(66)$$
$$\qquad\qquad\quad SU(2)\ \ U(1)$$

W at 100 GeV makes force effectively weaker.

Start of universe?

The SU(3) interquark force is of the same strength as U(1) × SU(2).

(b)

(c)

FIGURE 13.1 Weak electromagnetic and strong forces change strengths at high energy. (a) Summarises the data as they were around 1985. (b) Following LEP, the precision was such that it is now clear that the simplest extrapolation causes the strengths of the forces to miss slightly. (c) If SUSY particles occur at around 1-TeV energy scales, they will affect the extrapolation in such a way that the forces may indeed be united at very high energies.

Table 13.2 Do Exotic Forces Operate at Very Short Distances?

The nuclear force is strong but of such short range that its effects are feeble at atomic dimensions. We would be unaware of this force were we unable to probe distances of 10^{-15} m. The "weak" force also is very short range and only reveals its full strength when distances of $\sim 10^{-17}$ m are probed. The energies required to do this are $\sim 100\,\text{GeV}$ and under these conditions we have discovered that the "weak" force is really effectively as strong as the electromagnetic force. It is only when studied from afar that they appeared to be more feeble, hence, "weak."

This raises the question of whether there are other forces, transmitted by bosons, whose masses are far in excess of what present experiments can produce, and whose effects are limited to exceedingly short distances. Some Grand Unified Theories predict that there are such forces acting over a distance of $\sim 10^{-30}$ m that can change quarks into leptons, and, hence, cause a proton to decay. Although such forces are beyond our ability to sense directly at presently available energies, their effects might still be detected. This is because quantum mechanics comes to our aid.

When we say that the proton is 1 fm in size, we are making a statement of probability. About once in 10^{30} years, according to such theories, quarks in the proton will find themselves within 10^{-30} m of one another and then this force can act on them, causing the proton to decay. Thus we can study very short distance phenomena if we have great patience and wait for the "once in a blue moon" configuration to happen. With enough protons we can tip the odds in our favour and need not wait so long. If we have 10^{30} protons, as in a huge container of pure water, there may be one or two protons that decay each year. The challenge is then one of seeing this and ensuring that it can be distinguished from other natural background effects such as radioactivity and cosmic rays. Definite proof of proton decay is still awaited.

SUPERSYMMETRY

That the electromagnetic, weak and strong forces could change their characters and have similar strengths at extremely high energy (Figure 13.1) was first realised in the 1970s, and experiments confirmed this tendency. As we have seen, data from LEP showed that while for QED α is approximately $1/137$ at low energy, it grows to about $1/128$ by the energy at which the Z^0 begins to manifest its role. The strength of the analogous quantity in QCD, α_s had fallen from a value of 0.3 at the energies of a few GeV where the ψ and charmonium are prominent, to about 0.1 at the 90 GeV where the Z^0 is produced. Extrapolating these trends to higher energies showed that they could merge at energies around 10^{15} GeV.

At least that is how it appeared at first. However, by the time LEP had completed its decade of measurements, the data had become precise enough to show that the extrapolation did not bring all these strengths to a single value at the same energy. It was tantalising: from what had initially been very different magnitudes, they headed towards one another over vast ranges of energy, but then just missed very slightly.

Given that the extrapolation had been made over 15 orders of magnitude, it was actually remarkable that they merged so well at all, as this assumed that there are no other massive particles that could have contributed to quantum fluctuations. Any massive particle and its antiparticle could contribute and change the evolution of the various α once the energy exceeds that required to create the new massive particles. The interesting thing is that such heavy particles have been predicted to exist — known as supersymmetric particles — and when their effects are included in the evolution, it subtly focuses the changing values such that all the α merge into neat coincidence. This does not of itself prove that supersymmetry — SUSY — is real, but it is tantalising. So what is SUSY?

The particles that we have met fall into two classes according to the amount of spin. Bosons have spins that are zero or integer multiples of Planck's quantum; fermions have spins that are half-integers (1/2, 3/2, etc.). Fermions obey the Pauli exclusion principle which restricts their combinations; thus for electrons, it gives rise to the patterns of the periodic table; for protons and neutrons, it determines which atomic nuclei are stable and common or unstable and rare; for quarks, it determines the patterns of the Eightfold Way. Bosons, by contrast, are like penguins — the more the merrier; there are no such restrictions on the photons that form intense beams of laser light. Bosons, such as photons, W and Z bosons, and the gluons, are the carriers of the forces that act on the "matter" particles, the fermions. We have seen that there is a tantalising symmetry between the leptons and the quarks in the fermion sector; on the other hand, we have discerned a clear commonality among the force-carrying bosons. As the forces are transmitted by the particles and, in turn, the particles feel forces, a natural question is whether the particles of matter — the leptons and quarks — are related to the force carriers. The premise of SUSY is that they are.

We could at this point just assert that this is so and describe what happens, but the idea is rooted in profound ideas about which some readers might wish to know. If not, then omit the next paragraph.

All known laws of physics remain true when particles are moved from one point to another (invariance under "translation"), rotated in space, or "boosted" (have their velocity changed). These invariances are known as symmetries, the full set of symmetries in space and time that transform a particle in some state into the identical particle in some other state (such as position or velocity) being known as Poincare symmetry (after the French mathematician). In the 1970s, physicists discovered that there is another symmetry that is allowed within the relativistic picture of space-time and quantum mechanics: the loophole that takes us beyond the Poincare symmetries is that the new symmetry changes the particle's spin by an amount 1/2, while leaving other properties, such as mass and electric charge, unchanged. This symmetry is known as supersymmetry, and the action of changing spin by 1/2 means that it transforms a boson into a fermion (or vice versa) with the same mass and charge. It can be proved mathematically (but it takes an advanced course in mathematical physics to do so, so please take it on trust) that supersymmetry is the maximum possible extension of Poincare symmetry. Thus the discovery of supersymmetry in Nature would be profound in that we would have finally identified all of the symmetries that space-time has.

If supersymmetry is realised in Nature, then to every boson there exists a fermion with identical mass and charge; and to every boson there exists such a fermion. Clearly this is not the case in reality: there is no electrically charged boson with the mass of the electron, no massless coloured fermion sibling to the gluon, and so forth. So with apparent evidence already ruling against SUSY, why is there so much interest in it?

If our understanding of the symmetries in space and time that arise in relativity and quantum mechanics are correct, then supersymmetry seems unavoidable. As there is no obvious sign of it, then either our fundamental understanding of Nature is wrong in some respect, which could have itself profound implications, or SUSY is indeed at work but its effects are hidden. The suspicion is that the latter is the case and that SUSY is an example of what is known as a "hidden" or "spontaneously broken" symmetry (this concept is described in the section on the Higgs boson). Most theorists suspect the superpartners of the known particles do have the same charges as their siblings but are much more massive. Thus it is possible that SUSY is a symmetry in all but the masses of the particles. Such an idea is not outlandish in that we already have examples with the known particles and forces of symmetries that are spoiled by mass. We have seen that the W and Z are siblings of the photon and that the weak force and electromagnetic interactions are also siblings: in all cases, it is the mass of the particles that masks the symmetry. The idea that mass has hidden the fundamental symmetries is now a paradigm that drives current research; proving it will require the discovery of the Higgs boson and of massive supersymmetric particles.

In supersymmetry the families of bosons that twin the known quarks and leptons are "superquarks" (known as "squarks") and superleptons ("sleptons"). If SUSY were an exact symmetry, each variety of lepton or quark would have the same mass as its slepton or squark sibling. The electron and selectron would have the same mass as one another; similarly, the up quark and the "sup" squark would weigh the same, and so on. In reality this is not how things are. The selectron cannot be lighter than 100 GeV; otherwise it would have shown up in experiments. Thus either it does not exist, and SUSY is not part of Nature's scheme, or the selectron is hundreds of thousands of times more massive than the electron. Similar remarks can be made for all of the sleptons or squarks.

An analogous statement can be made about the super-partners of the known bosons. The naming pattern for the superfermions that partner the known bosons is to add the appendage "-ino" to denote the super-fermion partner of a standard bo-son. Thus there should exist the photino, gluino, zino, and wino (the "ino" pronounced eeno; thus for example, it is weeno and not whine-o). The hypothetical graviton, the carrier of gravity, is predicted to have a partner, the gravitino. Here again, were su-persymmetry perfect, the photino, gluino, and gravitino would be massless, like their photon gluon and graviton siblings; the wino and zino having masses of 80 and 90 GeV like the W and Z. But as was the case above, here again the "inos" have masses far greater than their conventional counterparts.

Without SUSY, attempts to construct unified theories tend to lead to nonsensical results, such as that certain events could occur with an infinite probability. However, quantum fluctuations, where particles and antiparticles can fleetingly emerge from the vacuum before disappearing again, can be sensitive to the SUSY particles as well as to the known menu. Without the SUSY contributions, some calculations give infinite nonsense; upon including the SUSY contributions, sensible results emerge. The fact that the nonsensical results have disappeared when SUSY is at work encourages hope that SUSY is indeed involved in Nature's scheme. Getting rid of nonsense is, of course, necessary, but we still do not know if the sensible results are identical with how Nature actually behaves. So we have at best indirect hints that SUSY is at work, albeit behind the scenes at present. The challenge is to produce SUSY particles in experiments, thereby proving the theory and enabling understanding to emerge from the study of their properties.

SUSY might be responsible for at least some of the dark matter that seems to dominate the material universe (p. 193). From the motions of the galaxies and other measurements of the cosmos, it can be inferred that perhaps as much as 90% of the universe consists of massive "dark" matter or dark energy, dark in the sense that it does not shine, possibly because it is impervious to the electromagnetic force. In SUSY, if the lightest superparticles are electrically neutral, such as the photino or gluino, say, they could be metastable. As such, they could form large-scale clusters under their mutual gravitational attraction, analogous to the way that the familiar stars are initially formed. However, whereas stars made of conventional particles, and experiencing all the four forces, can undergo fusion and emit light, the neutral SUSY-inos would not. If and when SUSY particles are discovered, it will be fascinating to learn if the required neutral particles are indeed the lightest and have the required properties. If this should turn out to be so, then one will have a most beautiful convergence between the field of high-energy particle physics and that of the universe at large.

HIGGS BOSON

Throughout this story we have met examples of "nearly" symmetry: things that would have been symmetric were it not for the spoiling role of mass. The electroweak force, for example, appears as two distinct pieces in phenomena at low temperature or energy — the electromagnetic and the weak forces. We now know that the reason is because the former is transmitted by photons, which are massless, whereas the latter involve W and Z bosons, which are massive. It is their great masses that enfeeble the "weak" force. The W is at work in the transmutation of hydrogen into helium in the Sun; the Z determines some aspects of supernova explosions that pollute the cosmos with the elements needed for planets, plants, and living things. We are stardust, or if you are less romantic, nuclear waste. Had the W not been massive, the Sun would have expired long before life on Earth had begun. So we owe our existence, in part, to the massive W and Z bosons, in contrast to the massless photon. This is but one example of where a symmetry has been broken by mass.

We have seen also how each generation of quarks and leptons appear to behave the same in their responses to the forces; it is their different masses that distinguish them. And if the ideas of supersymmetry are correct, then here too it is mass that spoils the symmetry. Once more we are fortunate that nature is like this. In the universe, the lightest charged particles are electrons and quarks. These are fermions and obey the Pauli principle, which gives rise to structure, as in protons, atoms, and ultimately chemistry and biology. Had supersymmetry been exact, then electrically charged bosons, such as selectrons, would have been as light as the electrons, and with no Pauli principle to intervene, would have clustered together, destroying the structures on which we depend. Had the $W^\pm$ been massless, they too would have mutually attracted without penalty; this would have altered the atomic forces such that electrons would probably have been permanently entrapped within atoms much as quarks are within hadrons.

A universe with perfect symmetry would have been a peculiar place. Our existence seems to depend on symmetry having been spoiled by the property that we call mass.

So what is mass? Where does it come from? The mass of you and me is due primarily to the nuclei of atoms, which is dominantly due to the protons and neutrons within. Their mass comes from the energy associated with the fields that confine the quarks in their prisons. The masses of the quarks themselves, and of the leptons, arise from some other mechanism; it is about the origin of the mass of the fundamental particles that I address my question. The received wisdom is that it is a property of what is known as "hidden symmetry" or "spontaneous symmetry breaking." The theory is due to Peter Higgs and it has consequences that are only now becoming accessible to experimental test.

To learn more about this fascinating subject I refer you to *Lucifer's Legacy* (see bibliography). In brief, this is a phenomenon that is well known throughout Nature. At high energies, or warm temperatures, a symmetry may be manifested that becomes altered at lower temperatures. For example, the molecules of water are free to take up any direction within the warm liquid, but once frozen, they form beautiful structures, as exhibited by snowflakes. A pattern emerges in the cold even though the underlying laws showed no such prejudice for one direction over another.

Getting nearer to the Higgs idea we come to magnets. In iron, each electron is spinning and acts like a small magnet, the direction of its north–south magnetic axis being the same as its axis of spin. For a single isolated electron this could be any direction, but when in iron, neighbouring electrons prefer to spin in the same direction as one another, as this minimises their energy. To minimise the energy of the whole crowd, each little magnet points the same way and it is this that becomes the magnetic axis of the whole magnet. This is a beautiful example of broken symmetry, as the fundamental laws for spinning electrons have no preferred direction, whereas the magnet clearly does. Heat up the magnet beyond the critical temperature at which it undergoes "phase change;" the original symmetry is restored and the magnetism disappears. Instead of doing this, we could stay in the cool magnetic phase and give the electrons a small pulse of energy so that they wobble. It is possible to make the spins wobble, the direction of spin or local magnetic north varying from point to point in a wavelike fashion. These are known as spin waves; as electromagnetic waves are manifested by particle quanta, photons, so spin waves can be described as if bundled into "magnons."

Ideas such as these were developed by Philip Anderson around 1960. Higgs' theory is built along these lines and applied to the universe, built on the perception that the vacuum is really a structured medium and that the analogy of the magnons become particle manifestations known as Higgs bosons.

The basic principle is that physical systems, when left to their own devices, attain a state of lowest energy. To do so they may change their phase. Now imagine pumping a container empty and lowering its temperature as far as possible. You would expect that by doing so the volume in the container would have reached the lowest energy possible since by adding anything to it, you would be also adding the energy associated with that stuff. However, in the Higgs theory, the "Higgs field" has the bizarre effect that when you add it to the container, it *lowers* the energy still further. At least this is the case so long as the temperature is below some $10^{17°}$.

The symmetric universe was hotter than this only for a mere 10^{-14} seconds after the Big Bang, since the Higgs field has been locked into its fabric. When particles interact with this field, they gain inertia, mass. As the freezing of water selects six directions for the ice crystals to develop out of the rotational symmetry that had existed previously, so patterns emerge among the particles as the universe cools below the critical Higgs temperature. One aspect of these patterns is that the particles gain masses and hence their distinct identities.

These ideas are theoretical and will remain so until confirmed by experiment. In energy terms, $10^{17°}$ corresponds to some 1000 GeV (1 TeV). If the theory is correct, collisions among particles at such energies will create conditions whereby the full symmetry is revealed and the Higgs boson produced. We know that it must have a mass greater than 100 GeV as otherwise it would have been produced at LEP. At the other extreme, if it were greater than a few hundred GeV, it would affect precision measurements in ways that would have already been detected. That is how it is with the simplest realisation of Higgs' ideas. The concept of supersymmetry allows richer possibilities. There could be a family of Higgs bosons, the lightest of which could be around 100 GeV. To test these ideas by producing these particles, the LHC has been designed specifically to reach the energies that theory predicts are needed.

The relative masses of the fundamental particles and the strengths of the forces are delicately balanced such that life has been able to develop. We still do not know why this is, but Higgs' mechanism is believed to be crucial in having determined their properties. Discovering the Higgs boson(s), finding how many there are, and how they operate will thus be essential steps towards solving one of the greatest of mysteries.

WHY IS 1 TeV SPECIAL?

The plans to search for the Higgs boson have focussed on an energy scale of up to 1 TeV. This brief summary attempts to give a flavour of how this came about.

One way that mass affects the behaviour of particles is that it is possible to move faster than a massive particle but not a massless one. Only massless particles, like the photon, can travel at the speed of light; massive particles can be accelerated ever closer to Nature's speed limit but can never reach it. For example, the electrons at LEP got within 120 km/hour, and the protons at the LHC reach within 10 km/hour of light speed. This means that if a massive particle passes us heading south, say, we could (at least in our imagination) move ourselves southwards even faster, overtaking the particle such that we perceive it to be moving north relative to us. In the case of a particle that spins, like the electron, this has a profound consequence.

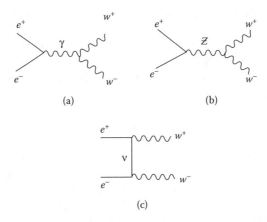

FIGURE 13.2 The most important Feynman diagrams that contribute to $e^+ e^- \rightarrow W^+ W^-$ according to the standard model of the electroweak interactions. These involve possible intermediate (a) photon, (b) Z^0, or (c) the exchange of a neutrino.

A spin 1/2 electron can have its spin oriented along the direction of its motion or opposite to it (p. 170). The former case is known as right-handed and the case of opposite orientation is called left-handed. Suppose that the electron is heading north and has its spin also pointing north: right-handed. Now consider what happens if we get into a fast moving vehicle that overtakes the electron and look back at it. If we are moving at a constant velocity this is known as an "inertial frame", so we will have viewed the electron in two different inertial frames: one when we were at rest and one when we were moving faster than the electron. In this latter case, when we look back on the electron we will perceive it as receding from us in a southerly direction; the direction of its spin however will appear to be unaltered and so the electron will appear to be left-handed. What was right-handed when viewed by us at rest becomes left-handed when viewed from a different inertial frame. The theory of relativity says that the laws of physics must appear the same to all experimenters moving at constant velocities relative to one another. The implication is that whereas for a massless particle, left- and right-handed states can act independently with different properties, for massive particles the interactions of left- and right-handed states are intimately linked. In the example of electron positron annihilation, if when at rest I see a right-handed positron annihilate with a left handed electron, an observer moving even faster than the electron will see both the electron and positron as right handed when they annihilate.

The electroweak theory obeys the constraints of relativity and indeed implies that electron and positron can annihilate in either of these combinations, right with right or right with left, and that this is also a consequence of them having a mass. Fundamental symmetries that underpin the Electroweak Theory and Standard Model imply that if the electron and positron were both massless, they would annihilate with their spins in the same direction: this corresponds to their total spin adding to one unit of $\hbar$. In the inertial frame where they collide from opposite directions, such as at LEP where one comes from the north and one from the south, this corresponds to one being right-handed and the other left-handed. For collisions at very high energy the electron and positron masses are negligible, but not exactly zero, so left and right become intertwined and there is a small chance for them to annihilate with both being left-handed or both right-handed. This situation enables their total spin to add to zero rather than one unit of $\hbar$.

This can only occur because their masses are not zero; but given that it does, we can ask — what happens next? One possibility is that a pair of W bosons emerge: W^+ and W^- with their total spins cancelling to zero. In particular, these massive W bosons, whose individual spins are one unit of $\hbar$, can have their spins oriented in any of $+1$, -1, or 0 along their directions of motion. The latter orientation creates a problem. When you calculate the chance for this to happen, the answer turns out to depend on the energy. As the energy grows, so does the chance for $e^+e^- \rightarrow W^+W^-$. Eventually the chance turns out to exceed 100%, which is physically nonsensical and highlights some incompleteness in the theory.

Everything is sensible if the electron mass is zero; but for a real electron with a finite mass, nonsense occurs. It seems unlikely that the theory simply is wrong as it has passed so many tests. Something extra seems to be demanded and this something must be related to the fact that the electron has a mass. This is where the Higgs boson plays a role.

The Higgs boson (H^0) has spin zero, is electrically neutral, and so can contribute to the process via $e^+e^- \rightarrow H^0 \rightarrow W^+W^-$. As the peak and trough of two waves can cancel, so can the wavelike nature of particles allow contributions among Feynman diagrams to cancel. So it is here. The quantum waves associated with $e^+e^- \rightarrow H^0 \rightarrow W^+W^-$ cancel the ones from $e^+e^- \rightarrow \gamma \rightarrow W^+W^-$ and $e^+e^- \rightarrow Z^0 \rightarrow W^+W^-$ and also that where a neutrino is exchanged between the e^+e^- (Figure 13.2). The net result is an answer that is both sensible (in that probabilities never exceed 100%) and consistent with the measurements at the presently available large, but finite, energies at LEP.

In these experiments, the Higgs boson was too massive to be produced and existed fleetingly as a "virtual" quantum effect; even so, for the results to make sense, the Higgs cannot be too far away from the energies in the experiments and it was from such measurements that an upper limit on its mass, as being less than a few hundred GeV, can be deduced empirically. It also hints at the association of the Higgs boson with the source of the electron's mass: if there was no Higgs boson, no electron mass, and so no annihilation in the spin zero configuration, there would have been no nonsense. The fact that the electron has a mass, which allows the spin zero configuration to occur, leads to potential nonsense unless the mechanism that gives that mass can also contribute to the $e^+e^- \rightarrow W^+W^-$ process and lead to a sensible total answer. Either this is the Higgs boson or it is something else that will be manifested on the TeV energy scale and awaits discovery at the LHC.

Even without the experimental results from LEP to guide us, it is possible to deduce an upper limit on the Higgs boson mass from the energy at which the nonsensical results would arise were it not present. The process $e^+e^- \rightarrow W^+W^-$ involves the weak bosons and so its strength, or chance, depends on the weak interaction strength.

This is controlled by Fermi's constant G_F, which he introduced in his original theory of β-decay (Chapter 8). The energy scale at which the Higgs must be manifested is:

$$\sqrt{\frac{8\pi\sqrt{2}}{3G_F}} \tag{13.1}$$

This equation implies that the energy would be bigger if the weak interaction were weaker (i.e., if G_F were smaller). This makes sense as if there were no weak interaction, the W^+W^- could not be produced and there would be no nonsense to be solved and hence no need for a Higgs all the way to infinite energy. For the real world, the weak interaction does exist and its strength is summarised by Fermi's constant with value $\sqrt{2}/G_F = 10^5$ GeV^2. Put that into the equation and you will find that the critical energy scale by which the Higgs must appear is about 1 TeV.

What if a Higgs boson does not exist? In this case the mathematical arguments summarised by these Feynman diagrams must break down and something else must happen. The loophole is that the mathematics assumes that the W and Z boson interactions remain at weak or electromagnetic strength even at extreme energies. That is what the standard model implies but suppose that they experience some new form of strong interaction, which cannot be described by the "perturbation theory" techniques that underlie the use of such Feynman diagrams. In this case the above arguments would fail. But the implication still remains: these new strong interactions will have to be manifested by the same 1 TeV scale. So either the Higgs boson exists or some weird new force operates. Either way, the TeV energy scale, which the LHC is the first to explore, will reveal physical phenomena that we have never seen before. Nature knows the answer; the LHC will reveal it.

14 Cosmology, Particle Physics, and the Big Bang

The universe has not always been as we find it now: very cold in deep space at a temperature of 3K (−270°C); about 300K here on Earth, warmed by sunlight; 10 million degrees at the centre of the Sun where nuclear fusion reactions power its furnace. Long ago the entire universe was like our sun, and even hotter. Collisions among particles in high-energy physics experiments reproduce some of the conditions of that early hot universe, and it is in that sense that the physics of the smallest particles and the cosmology of the universe at the largest scales come together.

The modern view of a large-scale universe consisting of island galaxies, each with billions of stars, originated with the American astronomer Edwin Hubble. In 1924 he showed that galaxies exist beyond our own Milky Way, a discovery that indicated the universe to be far, far greater than had been previously imagined, followed by the more astonishing discovery, in 1929, that these galaxies are all rushing away from one another. The further away from us a galaxy is, the faster it is receding, which shows that the universe is expanding in all directions.

As the pitch of sound rises and falls as its source rushes to or away from your ears, so the "pitch," or colour, of light emitted by a body changes when it is in motion relative to your eye. Coming towards you the light is bluer and when going away it is "red-shifted." By comparing the spectral lines of elements shining in remote galaxies with those measured in the laboratory, we can deduce their motion relative to Earth. The red shifts of remote galaxies showed Hubble that the universe is expanding, and more modern measurements of the red shifts show that for every million light years distance, the relative speed of separation is some 30 km per second. This expansion rate of 30 km s^{-1} for a separation distance of a million light years enables us to take a ratio which will have the dimensions of time:

$$\text{1 million light years/30 km s}^{-1} = \text{Time}$$

If you played the expansion in reverse, the galaxies would have been closer together than they are now. The amount of time given by the above arithmetic is how long ago the individual galaxies that we now see would have overlapped one another: gathered together in a single mass, their material would be compressed into an infinitesimal volume. This singular state of affairs is what we currently call 'the start of the universe' and we can estimate its age as follows.

One million light years is the distance that light moving at 3×10^5 km s^{-1} travels in 10^6 years. So the time is

$$\frac{3 \times 10^5 \text{ km s}^{-1} \times 10^6 \text{ years}}{30 \text{ km s}^{-1}} \equiv 10^{10} \text{ years}$$

Thus we get of the order of 10 billion years as the age scale of the observable universe. Of course, this is only a rough estimate. The expansion will not have always been uniform; there will have been some small change in the observed expansion rate as we peer deeper into the universe, or equivalently, back in time; how sure are we that this can be extrapolated all the way back to "creation?" The current best estimate for the age by this method is between 12 and 14 billion years.

If this was all that we had to go on, we would regard it as indicative but hardly a strong proof. It is when we compare this with other independent measures of age, and find similar results, that we become confident that this is some measure of "truth." In Chapter 3 we saw how the half-lives of radioactive atomic nuclei enable the age of the Earth and some stars to be estimated. This gives the oldest stars' ages that are between 9 and 15 billion years, which is consistent with the Hubble method, though less precise. Recently an age of 13.7 billion years, with an uncertainty of only 200 million (0.2 billion) years, has been inferred from measurements using the microwave background radiation.

As seen in Chapter 1, in 1965, Arnold Penzias and Robert Wilson discovered black-body radiation whose intensity is the same in all directions in space, implying that it has to do with the Universe as a whole. The existence of this uniform thermal radiation existing throughout an expanding universe teaches us a lot. As the universe expands, so the thermal radiation will cool down, its absolute temperature halving as the radius doubles. Conversely, in the past when the universe was much smaller, the thermal radiation was correspondingly hotter. Its present temperature is $3°$ above absolute zero; extrapolating back 10 thousand million years reveals that the early, extremely dense universe was incredibly hot, which all agrees with the 'Hot Big Bang' model of the birth of the universe that had been suggested as early as 1948 by Alpher, Gamow, and Herman.

In the past 20 years, this microwave background has been studied using instruments on satellites. These show that its temperature is 2.725 K (above absolute zero) to an accuracy of 1 in 1000. However, the most recent data do even better than this and subtle fluctuations have begun to emerge.

In 1992, NASA's Cosmic Background Explorer, known as COBE, found deviations at the level of 1 part in 10,000 relative to the 2.725. These are related to variations in the density of matter in the early universe, from which the first galaxies formed. COBE's results were remarkable, but its resolution — the ability to discriminate from one point to another — was poor. If you hold both hands at arms length, that is the minimum angular size that COBE could resolve: about 15 times larger than the Moon's apparent size. Inspired by this discovery, in 2001 the Wilkinson Microwave Anisotropy Probe, known more easily by its acronym WMAP, was launched. This has far better resolution and has measured the fluctuations in greater detail (see Figure 14.1). From these measurements it is possible to determine the size of the universe at the time the first galaxies were forming and the time that light has travelled since that epoch. This gives us the age of the universe since the emergence of galactic matter to an accuracy of about 1%: 13.7 billion years.

So we live not in a static unchanging universe, but one that is developing in time. The realisation that the infinite variety in our present universe evolved from an early cauldron provided a totally new perspective on astrophysics and high-energy physics in the latter years of the 20th century. Much of the diversity in the laws governing the forces that act within and between atoms in matter as we know it today has arisen as a result of the changes that occurred in the universe during its early moments.

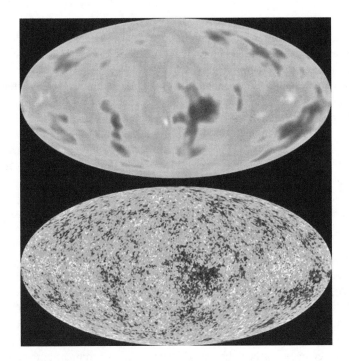

FIGURE 14.1 COBE and WMAP images of the microwave background sky. The upper and lower images show, respectively, the COBE (1992) discovery of fluctuations in the temperature of the microwave background radiation, and the improved resolution that resulted from data with the WMAP in 2003. (Source: NASA/WMAP Science Team.)

The received wisdom is that our long-standing experience, which has been gleaned from experiments restricted (we now realise) to "low" energies, may be only a partial guide in constructing theories of the hot early universe. Ongoing attempts to construct theories that go beyond the Standard Model uniformly suggest a new-born universe in which exotic processes were commonplace but which have been long since hidden from view. These new lines of enquiry have provided fresh insights into the workings of the early universe and have brought the previously disparate subjects of high-energy physics and cosmology together.

They have also revealed that the stuff that we are made of is but a flotsam on a sea of dark matter. And most recently, evidence has emerged for a mysterious "dark energy" pervading the cosmos. To appreciate these new discoveries we need first to recall some history.

DARK ENERGY

The first modern description of gravity in space and time was Einstein's theory of general relativity. Widely regarded as the pinnacle of human intellectual achievement in the 20th, or arguably any, century, it troubled Einstein: he thought the universe was static whereas his general theory of relativity in its original form dictated otherwise. To obtain a stable solution of the gravitational field equation that would lead to a static universe, he added a "cosmological constant" to the theory, which in effect counterbalanced the attraction of gravity.

However, this recipe did not actually do what he had hoped for: Einstein's *static universe* would actually be unstable because the uneven distribution of matter in it would ultimately tip the universe into either a runaway expansion or to collapse. He called it his greatest blunder. When Edwin Hubble later showed that the universe is actually expanding and not static at all, Einstein's cosmological constant became largely forgotten. It is ironic that recent data suggest that something like the cosmological constant is needed in Einstein's theory — not to stabilise the universe, for that it cannot, but to vary the rate of expansion in accord with what is seen.

Observations from the COBE satellite on the formation of galaxies in the early universe, and the realisation that disconnected parts of the universe now seem to have been causally connected in the past, led to ideas that there may have been a period of rapid "inflation" in the instant of the Big Bang. This is theoretical speculation at present, but is consistent with the latest data from the WMAP satellite. However, what is certain is that following this the universe expanded for some 10 billion years, the rate of expansion slowing due to the inward pull of its own gravity. But then something happened; it seems that the rate of expansion began to increase.

The precision data from observatories on satellites show us how things are (were!) deep into the cosmos: far back in time (Figure 14.2). By comparing the red shifts of the galaxies with the brightness of the supernovae in them, we can build up a picture of how the universe has behaved at different stages in its history. This has recently shown that the rate of expansion is increasing.

Initially the expansion seems to have slowed under the gravitational tug of the matter in the universe. This was the situation for up to about 10 billion years; but during the past 5 billion years, the expansion seems to have begun to accelerate. It is as if there is some repulsive force at work, what science fiction afficianados would perhaps term *anti-gravity*. Why did it only turn on 5 billion years ago?

Ironically a possible explanation lies in resurrecting Einstein's "greatest blunder," the cosmological constant, in a new guise. This would correspond to there being an all-pervading form of "dark energy" in the universe which has a *negative* pressure.

The reason for the change from deceleration to accelerating expansion is as follows. If the volume of the universe doubles, the density of matter halves and the gravitational drag falls as a result. If dark energy is described by the cosmological constant, then it remains unchanged. Thus the density of matter in an expanding universe disappears more quickly than dark energy. Eventually the effects of dark energy win and its repulsive expansion begins to dominate over the attractive slowing.

If this scenario continues into the far future, dark energy will ultimately tear apart all gravitationally bound structures, including galaxies and solar systems, and eventually overcome the electrical and nuclear forces to tear apart atoms themselves, ending the universe in what has been called a "Big Rip." On the other hand, dark energy might dissipate with time, or even become attractive. Such uncertainties leave open the possibility that gravity might yet rule the day and lead to a universe that contracts

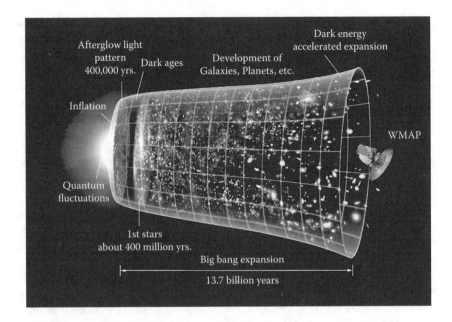

FIGURE 14.2 History of the universe. The 400,000-year-old afterglow of the Big Bang contained fluctuations, revealed by COBE and WMAP, that would eventually seed the galaxies of stars. The first stars appeared only after about 400 million years. An accelerated expansion seems to have taken over during the past 5 billion years, perhaps due to dark energy. The period before the afterglow can be studied by high-energy particle physics experiments. (*Source*: NASA/WMAP Science Team.)

in on itself in a "Big Crunch." While these ideas are not supported by observations, they are not ruled out. Measurements of acceleration are crucial to determining the ultimate fate of the universe in Big Bang theory.

So the current evidence suggests that the universe is filled with a mysterious dark energy, which might be a manifestation of something like the cosmological constant. At the moment no one knows precisely what it is. In addition to dark energy, there are hints of other stuff lurking in the dark: "dark matter." The answers to what this consists of may lie in cosmology or in particle physics, or both. The present situation is the subject of the next section.

DARK MATTER

The overall red shifts in the light emitted by stars in a galaxy teaches us about the motion of a galaxy as a whole. Within any galaxy the relative amounts of the shifts reveal the speeds of stars relative to the centre. This shows that most stars in spiral galaxies orbit at roughly the same speed, which is quite different to what we expect. Spiral galaxies appear to have a lot of mass in the centre such that the outer stars are spiralling around the centre, held in place by gravity. Orbiting around a central mass should give the outer stars a slower speed than the near-in ones; for example, like the planets orbiting our Sun where near-in Mercury moves faster than Earth, whereas the outer planets move more slowly than we. This is known as the galaxy rotation problem. Either Newtonian gravity does not apply universally or there is an extra gravitational tug due to considerably more matter existing in and around the galaxies than we see. This invisible stuff is known as *dark matter*.

Much of the evidence for dark matter comes from the study of the motions of galaxies. Galactic rotation curves, which illustrate the velocity of rotation versus the distance from the galactic center, suggest that galaxies contain a roughly spherical halo of dark matter with the visible matter concentrated in a disc at the center. Our Milky Way galaxy appears to have roughly ten times as much dark matter as ordinary matter.

Overall, dark matter has vastly more mass than the "visible" part of the universe. Only about 4% of the total mass in the universe can be seen directly. In short: 96% of the universe is missing, 74% being dark energy with dark matter comprising the remaining 22%.

Dark matter is classified as cold or hot. *Hot dark matter* consists of particles that travel near the speed of light, and as such have only small masses. One kind of hot dark matter is known, the neutrino, but the three known varieties of neutrinos make only a small contribution to the density of the whole. Furthermore, hot dark matter cannot explain how individual galaxies formed from the Big Bang. The COBE and WMAP measurements of the microwave background radiation indicate that matter has clumped on very small scales, and slow-moving particles are needed for this to happen. So although hot dark matter certainly exists in the form of neutrinos, they can only be part of the story, and to explain structure in the universe we need cold dark matter.

It seems that the smallest structures, such as stars, formed first, followed by galaxies and then clusters of galaxies. This model of structure formation requires *cold dark matter* to succeed: ordinary matter made of quarks in protons and other baryons left over from the Big Bang would have had too high a temperature to collapse and form smaller structures, such as stars.

At present, the density of ordinary baryons in the universe is equivalent to about one hydrogen atom per cubic meter of space. There may be clumps of hard-to-detect conventional matter, such as objects like Jupiter that do not burst into light, but are still detectable by their warmth, infra-red radiation; however, they can only contribute a small amount of the required amount of dark matter.

The current opinion is that that dark matter is made of one or more varieties of basic particles other than the usual electrons, protons, neutrons, and ordinary neutrinos. Supersymmetry contains massive neutral particles, such as photinos and gluinos, the lightest such "neutralino" having the possibility of being metastable and as such a candidate for seeding dark matter. Another candidate is a "sterile neutrino," which has been invoked by some theorists to help explain the anomalously light masses of the known neutrinos.

The need for cold dark matter is a potentially cosmological pointer to physics beyond the Standard Model. Finding supersymmetry or sterile neutrinos at the LHC could make a profound connection between cosmology and particle physics.

THE UNIVERSE TODAY

If the 14 billion years of the universe were likened to 14 hours of the day, all of our scientific research would have been performed during the span of a single heartbeat. Astronomers study distant galaxies whose light started its journey to Earth millions of years ago. From these observations we can, in effect, study the universe of long ago and check that the laws of physics applicable today are the same as those operating in that past.

However, we cannot see back into the first 400,000 years of the universe by observations with optical or radio telescopes, or by other electromagnetic means. The background radiation originated in that epoch and is a veil beyond which our eyes cannot see. Nonetheless, there are relics of earlier epochs such as the abundance of helium and other light elements relative to hydrogen in the universe, the density of matter relative to radiation photons and, closest to home, the existence of us: matter to the exclusion of antimatter. These are the observable signals, the beacons that show what occurred in that hidden first 400,000 years. Clues as to why they are as they are have emerged from high-energy particle collisions where, in the laboratory, local hot conditions are created that imitate the state of that early universe.

Most of the matter in the universe as viewed today is in stars that shine by burning light elements and making heavier ones. These processes have been well understood for over 50 years, and calculations imply that all of the nuclei of the atoms heavier than lithium, in particular the carbon in our bodies, the air in our lungs, and the Earth beneath our feet, were cooked in the nuclear furnaces of stars relatively recently.

These stars are collected into galaxies which in turn are gathered by gravity into clusters. Hubble's observations showed that these galactic clusters are moving apart from one another at a rate of about 30 kilometres per second for every million light years that they are currently separated. The known matter density is some two to ten times too low to provide enough gravitational pull to slow the universe's expansion to a point where it will eventually collapse back inwards. Dark matter and dark energy confuse the issue, and the long-term future of the universe is still undetermined.

Even though the properties of dark matter and dark energy remain poorly understood, there is no doubt that the number of baryons is swamped by the number of photons in the background radiation. The number density of neutrons and protons in the universe as a whole is about 10^{-6}/cm^3, whereas the density of photons in the background radiation is of the order of 500/cm^3, leading to a very small ratio of baryons to photons:

$$\frac{N_B}{N_\gamma} \sim 10^{-9\pm1}$$

This number has long been a puzzle. One's immediate guesses would suggest a ratio either near unity or much nearer zero. If the universe contains matter and no antimatter, then why are photons 10^9 times more abundant than baryons? The rational explanation would be that they were produced from annihilation of matter and antimatter. In the primordial fireball of the Big Bang, the energy density was so great that matter and antimatter were continuously created and destroyed. Why did the destruction leave such a 'small' excess of photons and leave an excess of matter? The origin of the asymmetry between matter and antimatter is one of the major unsolved puzzles in current particle physics. The existence of three generations of quarks and leptons allows a natural imbalance between matter and antimatter, as we saw in Chapter 11, which is realised empirically. While this arcane asymmetry in exotic strange and bottom matter is realised, it does not seem to solve the central problem.

At CERN, positrons and antiprotons have been created and then fused together to make atoms of antihydrogen. The aim is to capture them in large enough amounts that a spectrum of antihydrogen can be excited and measured. It would then be compared with the analogous spectrum of hydrogen. Any difference could show a measurable difference between hydrogen, the commonest element of our universe, and antihydrogen, that of the antiuniverse. However, no differences have yet been found. The mystery of the asymmetry between matter and antimatter at large remains unsolved.

FIVE STAGES IN A COOLING UNIVERSE

In the history of the universe, summarised in Table 14.1, we can identify five epochs of temperature. The first is the ultra hot of 10^{32} K during the first 10^{-43} seconds where quantum gravity effects are strong. We have little idea how to describe this epoch mathematically.

Table 14.1 Important Dates in the History of the Universe

Time	Temperature	Typical Energy	Possible Phenomena
10^{-43} s	10^{32} K	10^{19} GeV	Gravity is strong. Quantum gravity theory required (not yet available).
10^{-37} s	$> 10^{29}$ K	$> 10^{16}$ GeV	Strong, electromagnetic, and weak forces united.
10^{-33} s	10^{27} K	10^{14} GeV	Processes controlled by superheavy bosons start to freeze. Predominance of matter over antimatter probably established.
10^{-9} s	10^{15} K	10^2 GeV	Production of W bosons starts to freeze. Weak interaction weakens relative to electromagnetism. (Maximum energies attainable by latest high energy accelerators.)
10^{-2} s	10^{13} K	1 GeV	Colour forces acting on quarks and gluons, cluster them into 'white' hadrons. Colour forces hidden from view. Protons and neutrons appear.
100 s	10^9 K	10^{-4} GeV $\frac{1}{10}$ MeV	Nucleosynthesis—helium and deuterium created.
10^6 Years	10^3 K	1/10 eV	Photons decouple from matter. Background radiation originates (optical and electromagnetic astronomy cannot see back beyond this epoch).
10^{10} Years	3 K	10^{-3} eV	Today. Galaxies and life exist. Background radiation. Stars provide local hotspots.

Einstein showed in his general theory of relativity that space, time, and gravitational forces are profoundly related. In consequence, it is not clear what 'time' means when gravitational forces are singularly strong. Thus the notion of the 'start of the universe' is ill defined. We can extrapolate back from the present time to the point where the above unknowns appear; then if we naively project back still further we find that time zero was 10^{-43} seconds earlier. However, it is not clear what, if anything, this has to do with the 'start of the universe.' With this caveat applied to the first entry in Table 14.1, we can redefine all subsequent times to date from 'time that has elapsed since gravitational forces weakened.'

There is speculation that *superstring theory* may eventually give insights into the physics of this era. However, this is purely conjectural and there is as yet no evidence to show that superstring theory has anything to do with the physics of elementary particles. There is also much discussion about ideas of "inflation," which posit that the universe went through a sudden expansion in this initial period before settling in the gentler expansion that Hubble first detected. In the universe that we observe, there are regions that would appear never to have been causally connected, in the sense that they are separated by a distance larger than light could have travelled in 14 billion years. Yet the physics of these disconnected regions appears to be the same. If the universe underwent a sudden inflation initially, such that all of the observable universe originated in a small causally connected region, this paradox can be avoided. However, whether this demands inflation as the explanation remains an open question. There is no clear evidence to demand the idea, but the theory does have testable consequences about the nature of matter in the present universe that can be confronted with data from the COBE and WMAP satellites. As such, it is possible to rule out the theory, which makes it useful for science. However, although such ideas are written about regularly in popular science magazines, and have a "gee whiz" excitement value, and may even turn out to be true, they are not established at anything like the level of the material that appears in the rest of this book.

The hot epoch is a period of isotropy and no structure. At temperatures above 10^{29}K, the typical energies of over 10^{16} GeV are so great that the fireball can create massive particles, such as W, Z, and top quarks as easily as photons, light quarks, leptons, and their antiparticles. The strong, weak, and electromagnetic interactions are believed to have similar strengths, quarks and leptons readily transmute back and forth, and a unity abounds that will soon be lost forever.

The warm epoch covers 10^{29} to 10^{15}K, where typical energies are 10^{16} to 100 GeV. This period of cooling is the one where first the strong and then weak and electromagnetic interactions separate and take on their individual characters.

In matter today, quarks and gluons are "frozen" into the clusters that we know as protons, neutrons, and other hadrons. According to theory, at higher temperatures they would have been "melted" into a quark gluon plasma (QGP). This state of matter is named by analogy with the familiar electrical plasma, such as found in the Sun. At low temperatures on Earth, electrons and protons are bound in atoms; at temperatures above about 200,000K, or energies above 15eV, not even hydrogen atoms survive and the electrons and protons roam free as independent electrically charged "gases," a state of matter known as *plasma*. Quarks and gluons are colour-charged; bound into white hadrons at low temperatures, the QGP is predicted to be a complex coloured analogue of familiar plasma. Experiments at RHIC — the Relativistic Heavy Ion Collider in New York — and at CERN show the first hints of a phase transition from hadrons to QGP. By 2010 we can expect quantitative measurements to become available on the behaviour of QGP and thereby test if our ideas on the evolution of hadronic matter are correct.

The lower end of this energy range became accessible to LEP in the 1990s. The symmetry between the weak and electromagnetic interactions is only broken at the very cold end of this epoch, so its restoration in the laboratory has been a profound confirmation of the idea that we live in the cold asymmetric aftermath of a once hot symmetric creation. The discoveries of the W and Z were major steps in this regard; the discovery of the Higgs boson at the hotter extent of this region is now a pivotal requirement to reveal the deeper symmetry.

Next we come to the cool epoch. It is now so cool relative to 100 GeV that the 'weak' force has indeed become weak. Electric charge and colour are the only symmetries remaining in particle interactions.

The physics of this cool epoch was studied in high-energy physics experiments during the latter half of the 20th century. At the cold end of this region the coloured objects have clustered into white hadrons and are hidden from view. We are blind to the colour symmetry and the strong interactions take on the form that binds the atomic nucleus. Temperatures of 10^7K, or in energy terms MeV, is where atomic nuclei exist but atoms are disrupted. This is the province of experimental nuclear physics; in the cosmos today, it is astrophysics — the physics of stars and the hot fusion that powers the centre of the Sun.

In the cool conditions today, a free neutron can decay $n \rightarrow pe^-\bar{\nu}$ but the reverse process $p \rightarrow ne^+\nu$ is trying to 'go uphill' and does not occur (for free protons). In the heat of the early Universe, electrons had enough thermal energy that they could collide with protons and cause the reverse process: $e^-p \rightarrow n\nu$. So there was a state of balance: neutrons were decaying $n \rightarrow pe^-\bar{\nu}$, producing protons and electrons, but the latter could bump into one another and revert back into neutrons: $e^-p \rightarrow n\nu$. The net number of neutrons and protons is preserved but many neutrinos are being produced.

As the temperature falls, the reverse reaction becomes increasingly difficult to achieve as there is no longer enough energy to make the more massive neutron from the initial ingredients. After about 1 microsecond, this reaction is frozen out and the whole process goes only one way: $n \rightarrow pe^-\bar{\nu}$. If this was the whole story, eventually all the neutrons would have disappeared, but they can bump into protons and bind to one another forming a stable nucleus of heavy hydrogen, the deuteron: $d \equiv np$. These deuterons in turn bump into further protons, gripping one another and forming the nuclei of helium, similar to what happens on the centre of the Sun today (see next section). This carried on until all the free neutrons were gone or until all the particles were so far apart in the expanding universe that they no longer bumped into one another.

The helium, deuterons, and a few light elements formed this way in the early universe would provide the fuel for the (yet to be formed) stars, in which they would be cooked to make the nuclei of yet heavier elements, feeding the periodic table of the elements. But first, let us see what became of those neutrinos.

One microsecond after the Big Bang, the neutrinos are free and are the first fossil relics of the creation. They move at high speed and, if massive, start clustering and contribute to the formation of galaxies. There are about 10^9 neutrinos for every atom of hydrogen today; so if a neutrino weighs more than one billionth of a proton, the totality of neutrinos will outweigh the known matter in the cosmos. Thus, quantifying neutrino masses is a big challenge with implications for our projections on the long-term future of the universe.

The universe expands and cools. In some ways it is like a huge piston filled with gas, in this case a gas of neutrinos. The rate of the expansion then depends on the pressure in the gas and its temperature. These depend among other things on the number of neutrinos in the gas volume, and in turn this depends on the number of distinct varieties.

If there are three varieties of light neutrinos, then the calculations imply that about 3 minutes after the Big Bang, matter consisted of 95% protons, 24 % helium nuclei, and small amounts of deuterium and free electrons. The helium abundance depends on the expansion rate and hence the number of varieties of neutrino: agreement with the observed helium abundance, the number above, fits best if there are three varieties of neutrinos. This number agrees with the "direct" measurement of neutrino species at LEP (p. 144). The abundance of deuterium depends on the density of "normal" matter. This value empirically suggests that this conventional matter is only a small part of the whole and as such adds credence to the belief in "dark matter." The precise role of dark matter and dark energy in the early universe is not well understood, not least because we do not yet know what the source and properties of these dark entities are. This is an exciting area of research, and discoveries at the LHC may begin to shed light on these issues.

After about 300,000 to 400,000 years, the temperature is below $10,000°$ and the mean energy per particle is below 10 eV. Electrons can combine with nuclei, forming electrically neutral atoms. Being neutral, these atoms have little propensity to scatter light, and the universe becomes transparent. This is how it is today: we can see with light that has travelled uninterrupted for millions, even billions of years.

This brings us to the cold epoch — the world of atoms, molecules, and the present-day conditions on Earth.

We now have the very antithesis of the isotropy present in the hot epoch. The different strengths in the natural forces and the various families of particles, nuclei, electrons, and neutrinos, upon which the forces act in differing ways, give rise to rich structure in the universe. Galaxies, stars, crystals, you and I, all exist.

LIGHT ELEMENTS AND NEUTRINOS

The nuclei of the heavy atomic elements that constitute the bulk of the material on Earth were formed inside stars. There, light nuclei fuse together producing heavier ones, in the process releasing the energy by which the stars are visible. If we can explain the existence of heavy elements in terms of light ones, then how were the light ones — hydrogen and helium — created?

The hydrogen and helium in the universe today were produced about 3 minutes after the Big Bang and are a relic of that time. The temperature then was about 10^9K and 'nucleosynthesis' (formation of nuclei) occurred very rapidly from the abundant neutrons and protons because these can fuse to form deuterium without being immediately ripped apart by hot photons: 'photodisintegration.' The temperature balance is very critical — cool enough that there is no photodisintegration, yet hot enough that two deuterons can collide violently and overcome their electromagnetic repulsion (they each carry positive charge due to their proton content).

All the neutrons form deuterium and all the deuterium forms helium-4:

$$n + p \rightarrow d + \gamma; \gamma + d \nrightarrow n + p$$
$$d + d \rightarrow^4 He + \gamma$$

Small amounts of helium-3 and lithium-7 were synthesised at the same time but production of heavier elements was prevented because there are no stable isotopes with atomic mass 5 or 8.

All the neutrons and a similar number of protons have formed light nuclei; excess protons remain as hydrogen. Thus the ratio of helium to hydrogen present today tells us the neutron–proton ratio at nucleosynthesis. Astrophysicists can compute this ratio and it turns out to be rather sensitive to the precise rate at which the universe was expanding. This is controlled by the density of matter and also by the number of light particles such as neutrinos. The observed ^4He abundance today includes a small amount that has been produced in stars subsequent to the first 3 minutes. Allowing for this, it seems that there can be at most threee or four varieties of neutrino in nature ($\nu_e, \nu_\mu, \nu_\tau, \ldots$).

We have seen how particle physics has uncovered a correspondence between lepton pairs and quark pairs ('generations'). At present we have evidence for possibly three generations: (up, down), (charm, strange), and (top, bottom) quarks, $(e, \nu_e); (\mu, \nu_\mu); (\tau, \nu_\tau)$ leptons. The helium abundance constraining the neutrino varieties to three or at most four implies that the three generations discovered are the sum total or that at most one more exists. The results from LEP on the lifetime of the Z^0 (p. 144) confirm that there can be only three. Thus do studies of the Big Bang constrain particle physics (and conversely).

QUARKSYNTHESIS AND THE PREDOMINANCE OF MATTER

The origin of heavy elements has been explained in terms of light ones. The light ones were formed by fusion of neutrons and protons, and these were formed by the clustering of coloured quarks into threes with no net colour. These quarks were produced in the original fireball where radiation energy was converted into quarks *and an equal number of antiquarks*. Thus the genesis of the ingredients of our present universe is understood, but how was this excess of quarks (and so matter) generated? Some asymmetry between matter and antimatter is required.

James Cronin and Val Fitch won the Nobel Prize for their part in the 1964 discovery that such matter–antimatter asymmetry does occur, at least in K meson decays. Starting with an equal mixture of K^0 and $\bar{K}^0$, their decays into $e^+\pi^-\bar{\nu}$ and $e^-\pi^+\nu$ are matter-antimatter correspondents since $e^+ \equiv e^-$ and $\pi^- = \pi^+$.

Yet the two sets of products are *not* equally produced: the decay $K^0 \rightarrow e^+\pi^-\bar{\nu}$ is about 7 parts in a 1000 more frequent than $\bar{K}^0 \rightarrow e^-\pi^+\nu$. In 2004, even more dramatic asymmetries were seen in the decays of B mesons and in the transmutations back and forth (known as "oscillations") of B^0 and $\bar{B}^0$ mesons.

These discoveries of asymmetric behaviour between matter and antimatter so far are only for the ephemeral flavours in the second and third generations. They do not (yet) teach us how the large-scale asymmetry between matter and antimatter in bulk arose.

Somehow all the antiquarks annihilated with quarks and all the antileptons annihilated with leptons leaving an excess of quarks and leptons. It is these "leftovers" that seeded our universe: most of the antimatter did not survive the first millionth of a second.

EXODUS

Will the universe expand forever, or is there enough matter in it to provide sufficient gravitational pull that it will eventually collapse under its own weight?

The first possibility presents a depressing outlook. Current attempts to unite quarks and leptons in grand unified theories imply that protons can decay such that all matter will decay with a half-life of about 10^{30} years. Thus all matter will erode away and the universe will end up as expanding cold radiation. The second possibility, collapse, could occur if neutrinos have enough mass.

In the primordial fireball, neutrinos should have been produced as copiously as were photons. Thus there should exist about a billion neutrinos per proton. Consequently, if an individual neutrino weighed as little as 1 to 10 eV, then the bulk of the mass in the universe would be carried by neutrinos, not baryons. As stars and other observed matter are built from the latter, there could be (at least) twice as much mass in the universe as we are currently aware of, and this could be sufficient to lead to ultimate gravitational collapse.

The discovery of neutrino oscillations shows that at least one and probably all of the three varieties of neutrinos have mass. However, no direct measure of this mass is yet available. All that we know for certain is that the *difference* in the masses is trifling, though less direct evidence suggests that each individually is very small.

This is in accord with computer simulations of how clusters of galaxies formed in the early universe. The best agreement with observations of how galaxies are distributed through space comes in models that have small neutrino masses. The "dark matter" seems to be a mix of hot (i.e., lightweight particles moving near the speed of light, probably neutrinos) and cold (massive potentially exotic stuff whose particle content is yet to be identified). Supersymmetry (SUSY) is a natural candidate for this, but until SUSY particles are discovered and their spectroscopy decoded, this must remain on the menu for the future.

Whether the universe will expand forever or eventually collapse in a Big Crunch is not yet certain. Will hints that the universe is now expanding faster than before ("dark energy") be confirmed, leading to a ripping a part of all matter in the far future? Why is the universe so critically balanced? Why are the masses of proton, neutron, electron, and neutrino so delicately tuned that we have evolved to ask these questions?

Such questions tantalise and may soon be answered. In turn, this will reveal new visions, leading us to ask and answer further questions that are still unimaginable.

EPILOGUE

It is little more than 100 years since Becquerel discovered the key that unlocked the secrets of the atom and led to our current all-embracing vision. Where will another century take us?

In 1983 when I wrote the original version of *The Cosmic Onion*, I concluded as follows:

"For all the wonders that our generation has been privileged to see, I cannot help but agree with Avvaiyar:

What we have learned
Is like a handful of earth;
What we have yet to learn
Is like the whole world"

A quarter of a century has elapsed since then. By focussing on concepts that were, in my opinion, established, I am pleased that almost nothing of that original *Cosmic Onion* has been proved to be so highly conjectural that it has been overthrown. Many new facts have come to be known, and much of the material in the latter part of this book bears testimony to that. Among them are several that were not anticipated.

The discoveries and precision measurements during the past 10 years have effectively established Electroweak Theory as a law of Nature. The discovery of the top quark completed three generations, but its mass being some 30 times greater than that of the bottom quark was a surprise. A question for the future is whether the top quark's huge mass is weird, or whether it is the only "normal" one, existing on the scale of electroweak symmetry breaking while all the other fundamental fermions are anomalously light. In any event, the top quark is utterly different, and when the LHC enables it to be studied in detail for the first time, further surprises could ensue. The top and bottom quarks turn out to be tightly linked in the weak decays, as summarised by the CKM theory, and almost disconnected from the other generations; this has caused the bottom meson to have an unexpected elongated lifetime that has enabled the study of $B-\bar{B}$ oscillations and the discovery that bottom meson decays give us a new arena for studying the violation of CP symmetry. Neutrino oscillations and their implication for neutrino masses were not on the theorists' agenda in 1983 but did emerge as a new research area at the start of the 21st century. And the precision data that probe the electroweak theory have revealed the quantum mechanical influence that the Higgs field has on the vacuum. In the cosmological arena we are now coming to the opinion that the universe is "flat," with appreciable dark matter (which some were beginning to speculate 20 years ago) and also dark energy (which was utterly unexpected). Quarks and leptons have been shown to be structureless particles to distances as small as 10^{-19} m, which is as small relative to a proton as the proton is to the dimensions of a hydrogen atom.

While these advances have been delivered by experiment, there were also theoretical concepts, such as superstrings, that in 1983 were only just about to emerge, and which are now a major mathematical research area. Large numbers of students who apply to do graduate studies in theoretical particle physics today cite superstrings as the area in which they wish to write their theses. It is possible that this will eventually prove to be the long sought Theory of Everything; and to read some popular science, or watch it on television, you might have the impression that it is established lore. However, there is at present no clear evidence to show that it has anything to do with the physics that has been revealed by experiment.

If superstring theory is the holy grail of physics, then a century from now its wonders will have been revealed and lucky those who will be around to know them. If it is not, then the Theory of Everything, if there is such, remains to be found. First though, one step at a time. The LHC is opening a new era in experimental exploration. If superstring theory is correct, the discovery of supersymmetry in experiments is a prerequisite. If the LHC reveals this, then we may expect to discover how it is that supersymmetry is not exact. With the discovery of the Higgs boson, or whatever should turn out to be the giver of mass to the particles, and an understanding of why supersymmetric particles have such large masses relative to those we know and love today, we may begin to get ideas on how to make a link between the ideal symmetric universe of superstring theory and the asymmetric cold universe that we have been restricted to so far. Or it may be that the answers lie in concepts that have not yet been written down. What distinguishes natural philosophy, or science, from other areas of thought is the experimental method. You can write down the most beautiful ideas, with the most elegant mathematics, such as the superstring theory for example. However, if experiment shows otherwise, your theories are useless. Time, and experiment, will tell — and that is for the future. For the present, that is why this book came to you with no strings attached.

Glossary

α:	*See* Coupling constant.
Alpha particle:	Two protons and two neutrons tightly bound together; emitted in some nuclear transmutations; nucleus of helium atom.
Angular momentum:	A property of rotary motion analogous to the more familiar concept of momentum in linear motion.
Antimatter:	For every variety of particle there exists an antiparticle with opposite properties, such as the sign of electrical charge. When particle and antiparticle meet, they can mutually annihilate and produce energy.
Anti(particle):	Antimatter version of a particle (e.g., antiquark, antiproton).
Atom:	System of electrons encircling a nucleus; smallest piece of an element that can still be identified as that element.
b:	Symbol for the "bottom meson."
***B*-factory:**	Accelerator designed to produce large numbers of particles containing bottom quarks or antiquarks.
Baryon:	Class of hadron; made of three quarks.
Beta-decay (beta-radioactivity):	Nuclear or particle transmutation caused by the weak force, causing the emission of a neutrino and an electron or positron.
Big Bang:	Galaxies are receding from one another; the universe is expanding. The Big Bang theory proposes that this expansion began around 14 billion years ago when the universe was in a state of enormous density and temperature.
Black-body radiation:	A hot black-body emits radiation with a characteristic distribution of wavelengths and energy densities. Any radiation having this characteristic distirbution is called black-body radiation. Any system in thermal equilibrium emits black-body radiation.

Boson:
Generic name for particles with integer amount of spin, measured in units of Planck's quantum; examples include carrier of forces, such as photon, gluon, W and Z bosons, and the (predicted) spinless Higgs boson.

Bottom(ness):
Property of hadrons containing bottom quarks or antiquarks.

Bottom quark:
Most massive example of quark with electric charge $-1/3$.

Bubble chamber:
Form of particle detector, now obsolete, revealing the flight paths of electrically charged particles by trails of bubbles.

CERN:
European Centre for Particle Physics, Geneva, Switzerland.

Charm quark:
Quark with electric charge $+2/3$; heavy version of the up quark but lighter than the top quark.

COBE:
Cosmic Background Explorer satellite.

Collider:
Particle accelerator where beams of particles moving in opposing directions meet head-on.

Colour:
Whimsical name given to property of quarks that is the source of the strong forces in QCD theory.

Conservation:
If the value of some property is unchanged throughout a reaction, the quantity is said to be conserved.

Cosmic rays:
High-energy particles and atomic nuclei coming from outer space.

Coupling constant:
Measure of strength of interaction between particles. In the case of the electromagnetic force, it is also known as the fine structure constant.

Cyclotron:
Early form of particle accelerator.

Down quark:
Lightest quark with electrical charge $-1/3$; constituent of proton and neutron.

Eightfold Way:
Classification scheme for elementary particles established *circa* 1960. Forerunner of quark model.

Electromagnetic radiation:
See Table 2.2.

Electron:
Lightweight electrically charged constituent of atoms.

Electroweak force:
Theory uniting the electromagnetic and weak forces.

eV (electronvolt):	Unit of energy; the amount of energy that an electron gains when accelerated by 1 volt.
$E = mc^2$ **(energy and mass units):**	Technically the unit of MeV or GeV, a measure of the rest energy, $E = mc^2$, of a particle, but it is often traditional to refer to this simply as mass, and to express masses in MeV or GeV.
Fermion:	Generic name for particle with half-integer amount of spin, measured in units of Planck's quantum. Examples are the quarks and leptons.
Feynman diagram:	Pictorial representatoin of particle interactions.
Fine structure constant:	*See* Coupling constant.
Fission:	Break-up of a large nucleus into smaller ones.
Flavour:	Generic name for the qualities that distinguish the various quarks (up, down, charm, strange, bottom, top) and leptons (electron, muon, tau, neutrinos); thus flavour includes electric charge and mass.
Fusion:	Combination of small nuclei into larger ones.
Gamma ray:	Photon; very high-energy electromagnetic radiation.
Gauge theories:	A class of theories of particles and the forces that act on them, of which quantum chromodynamics and the theory of the electroweak force are examples. The term "gauge" meaning "measure" was introduced by Hermann Weyl 70 years ago in connection with properties of the electromagnetic theory and is used today mainly for historical reasons.
Generation:	Quarks and leptons occur in three "generations." The first generation consists of the up and down quarks, the electron and a neutrino. The second generation contains the charm and strange quark, the muon and another neutrino; while the third, and most massive generation, contains the top and bottom quarks, the tau and a third variety of neutrino. We believe that there are no further examples of such generations.
GeV:	Unit of energy equivalent to a thousand million (10^9) eV (electron-volts).
Gluon:	Massless particles that grip quarks together making hadrons; carrier of the QCD forces.

Hadron: Particle made of quarks and/or antiquarks, which feels the strong interaction.

Higgs boson: Massive particle predicted to be the source of mass for particles such as the electron, quarks, W and Z bosons.

Ion: Atom carrying electric charge as a result of being stripped of one or more electrons (positive ion), or having an excess of electrons (negative ion).

Isotope: Nuclei of a given element but containing different numbers of neutrons.

J/ψ particle: Full name for the first discovered member of the charmonium family. Often referred to simply as ψ.

K (kaon):   Variety of strange meson.

keV: 1000 eV.

Kinetic energy: The energy of a body in motion.

LEP: Large Electron Positron collider at CERN.

Lepton: Particles such as electron and neutrino that do not feel the strong force and have spin 1/2.

LHC: Large Hadron Collider; accelerator at CERN.

Linac: Abbreviation for linear accelerator.

MACHO: Acronym for Massive Compact Halo Object.

Magnetic moment: Quantity that describes the reaction of a particle to the presence of a magnetic field.

Mass: The inertia of a particle or body, and a measure of resistance to acceleration; note that your "weight" is the force that gravity exerts on your mass so you have the same mass whether on Earth, on the Moon, or in space, even though you may be "weightless" out there.

Meson: Class of hadron; made of a single quark and an antiquark.

MeV: 10^6 eV.

meV: 10^{-6} eV.

Molecule: A cluster of atoms.

Microsecond: 10^{-6} seconds.

Muon: Heavier version of the electron.

Nanosecond: 10^{-9} seconds.

Neutrino: Electrically neutral particle; member of the lepton family; feels only the weak and gravitational forces.

Neutral current:	Weak interaction where no change takes place in the charges of the participants.
Neutron:	Electrically neutral partner of proton in atomic nucleus which helps stabilise the nucleus.
Neutrino:	Fundamental fermion; uncharged partners of electron, muon and tau.
Nucleon:	Generic name for neutron and proton, the constituents of a nucleus.
Nucleus:	Dense centre of an atom made of neutrons and protons. The latter give the nucleus a positive charge by which electrons are attracted and atoms formed.
Parity:	The operation of studying a system or sequence of events reflected in a mirror.
Periodic Table:	Table of the chemical elements exhibiting a pattern in the regular recurrences of similar chemical and physical properties.
Picosecond:	10^{-12} seconds.
Photon:	Massless particle that carries the electromagnetic force.
Pion:	The lightest example of a meson; made of an up and/or down flavour of quark and antiquark.
Planck's constant:	(h); a very small quantity that controls the workings of the universe at distances comparable to, or smaller than, the size of atoms. The fact that it is not zero is ultimately the reason why the size of an atom is not zero, why we cannot simultaneously know the position and speed of an atomic particle with perfect precision, and why the quantum world is so bizarre compared to our experiences in the world at large. The rate of spin of a particle also is proportional to h (technically, to units or half-integer units of h divided by 2π, which is denoted $\hbar$).
Positron:	Antiparticle of electron.
Proton:	Electrically charged constituent of atomic nucleus.
Psi:	A name by which the J or psi (ψ) meson is known.
QCD (quantum chromodynamics):	Theory of the strong force that acts on quarks.

QED (quantum electrodynamics): Theory of the electromagnetic force.

Quarks: Seeds of protons, neutrons, and hadrons (see pp. 79–82).

Radioactivity: *See* Beta-decay.

RHIC: Relativistic Heavy Ion Collider; accelerator at Brookhaven National Laboratory, New York.

SLAC: Stanford Linear Accelerator Center (California, USA).

SNO: Sudbury Neutrino Observatory, an underground laboratory in Sudbury, Ontario, Canada.

Spark chamber: Device for revealing passage of electrically charged particles.

Spin: Measure of rotary motion, or intrinsic angular momentum, of a particle. Measured in units of Planck's quantum.

Standard Model: The observation that fermions are leptons and quarks in three generations, interacting by forces described by QCD and the electroweak interactions, the latter being transmitted by spin 1 bosons: the photon, W and Z bosons, and gluons.

Strange particles: Particles containing one or more strange quarks or antiquarks.

Strange quark: Quark with electrical charge $-1/3$; more massive than the down quarks but lighter than the bottom quark.

Strangeness: Property possessed by all matter containing a strange quark or antiquark.

Strong force: Fundamental force, responsible for binding quarks and antiquarks to make hadrons, and gripping protons and neutrons in atomic nuclei; described by QCD theory.

SuperKamiokande: Underground detector of neutrinos and other particles from cosmic rays; located in Japan.

SUSY (supersymmetry): Theory uniting fermions and bosons, where every known particle is partnered by a, yet to be discovered, particle whose spin differs from it by $1/2$.

Symmetry: If a theory or process does not change when certain operations are performed on it, then we say that it possesses a symmetry with respect to those operations. For example, a circle remains unchanged after rotation or reflection; it therefore has rotational and reflection symmetry.

Synchrotron: Modern circular accelerator.

Tau: Heavier version of muon and electron.

Thermal equilibrium: The particles of a gas, for example, are in motion. Their average speed is a measure of the temperature of the gas as a whole, but any given particle could have a speed that is much less or greater than the average. Thus there is a range of speeds within the gas. If the number of particles entering a given range of speeds (more precisely, velocities) exactly balances with the number leaving, then the gas is said to be in thermal equilibrium.

Top quark: The most massive quark; has charge $+2/3$.

Uncertainty principle: One cannot measure both position and velocity (momentum) of a particle (or its energy at a given time) with perfect accuracy. The disturbance is so small that it can be ignored in the macroscopic world but is dramatic for the basic particles.

Unified Theories: Attempts to unite the theories of the strong, electromagnetic, and weak forces, and ultimately gravity.

Upsilon (γ):   Massive 9-GeV meson made from bottom quark and bottom antiquark.

Vector meson: A meson having one unit of spin, like the photon.

Virtual particle: A particle that is exchanged between other particles, transmitting a force or causing a decay. A virtual particle can exist fleetingly, apparently "borrowing" energy and momentum courtesy of the uncertainty principle.

Weak force: Fundamental force, responsible *inter alia* for beta-decay; transmitted by W or Z bosons.

W boson: Electrically charged massive particle, carrier of a form of the weak force; sibling of Z boson.

Weak interaction: A fundamental force of Nature. Most famous manifestation is in beta-decay. Also controls interactions of neutrinos.

Weinberg-Salam model: A name for the electroweak theory.

WIMP: Acronym for 'weakly interacting massive particle.'

WMAP: Wilkinson Microwave Anisotropy Probe.

Z boson: Electrically neutral massive particle, carrier of a form of the weak force; sibling of W boson.

Index